AN INQUIRY INTO THE NATURE OF REALITY

STRUCTURE, SCALE, AND THE ARCHITECTURE OF EXISTENCE

K V V S MURTHY

Developed In Dialogue With AI (ChatGPT)

notionpress.com

INDIA • SINGAPORE • MALAYSIA

ISBN
Paperback 979-8-90451-143-2
Hardcase 979-8-90451-144-9

Dedication

*To my wife, **Gowri Maheswari,** whose constant support and selfless sacrifice made this journey possible.*

"We have to remember that what we observe is not nature herself, but nature exposed to our method of questioning."

– Werner Heisenberg

Contents

Preface

This book did not begin with an answer.

It began with a question.

Or rather, with many questions—arising quietly, persistently, and sometimes uncomfortably—about space, time, matter, life, consciousness, and the strange coherence that seems to bind them all together.

I do not come to these questions as a physicist by training, nor as a philosopher by profession. My formal background lies in banking, freelance journalism, economics, finance and accounts, learning and development, strategy, and systems thinking—disciplines that, over decades, taught me something unexpectedly relevant: that complex realities rarely yield their meaning through isolated parts. They reveal themselves through **patterns, relationships, constraints, and emergent behaviour**.

That insight eventually led me beyond money, ledgers, balance sheets, and organisations, toward a deeper curiosity about reality itself.

The Spirit of This Inquiry

This book is not written to assert a worldview, defend a doctrine, or resolve metaphysical debates.

It is an **inquiry**.

An inquiry respects uncertainty. It proceeds step-by- step. It allows ideas to mature rather than forcing conclusions.

Where science speaks with evidence, it is honoured. Where science is silent, silence is respected. Where philosophy illuminates without replacing rigour, it is welcomed.

Throughout these pages, you will find questions posed before answers, structure before speculation, and coherence before certainty.

What This Book Is—and Is Not

This book is not:

a textbook in physics

a formal treatise in philosophy

a spiritual doctrine or manifesto

This book is:

an inquiry into the structure of reality

a reflection across physics, philosophy, and consciousness

an attempt to connect ideas across scales and disciplines

It is not written to convince. It is written to invite inquiry.

Here, questions matter more than conclusions.

It is an attempt to **connect domains often kept separate**—physics, information, life, intelligence, and reflection—without collapsing one into another.

The chapters move deliberately:

from **space and time**,

to **fields, symmetry, and mass**,

to **life, emergence, and intelligence**,

and finally, to **interdependence and meaning**.

Technical material is deliberately placed in **Appendices**, so that readers without formal scientific training can follow the main inquiry without interruption, while curious readers can explore deeper when they wish. The appendices are intentionally more extensive, reflecting the depth of material placed outside the main narrative.

A Note on Method

Two guiding principles shaped this work:

1. **Discipline of Boundaries** Physics is not metaphysics. Metaphor is not mechanism. Reflection is not proof.

Wherever these domains meet, they are allowed to **converse**, not replace one another. This inquiry seeks intuitive and conceptual understanding rather than mathematical derivation; wherever formal mathematics underlies the discussion, it remains implicit, with emphasis placed on meaning, structure, and coherence.

2. **Respect for Emergence** Some truths are not encoded upfront. They arise only through interaction, accumulation, and coherence.

This book itself evolved in that spirit.

This inquiry proceeds not by allegiance to any philosophical school, but by tracing relational structures and disciplined constraints through which reality becomes intelligible.

In Dialogue with AI: A Note of Acknowledgement

This book was **developed in dialogue with ChatGPT**, an artificial intelligence language model developed by OpenAI.

AI did not provide direct answers to the questions explored here. It did not supply authority, belief, or direction.

Instead, it served as:

a structured thinking partner,

a clarifier of arguments,

a mirror for coherence

and a disciplined editor for language and flow.

All responsibility for ideas, interpretations, and conclusions rests with the author.

How to Read This Book

This book does not demand linear reading.

Some readers may prefer to:

read the main Parts sequentially

skip Appendices initially

return later for technical clarification

Others may move back and forth.

Both approaches are valid.

The only request is this:

Read slowly, and allow questions to linger.

A Final Word Before We Begin

If this book has a quiet conviction, it is this:

Reality does not reveal itself all at once.
It yields insight through patience, coherence, and humility.

What follows is not a claim to truth,
but an invitation to inquiry.

Navi Mumbai, India **K V V S Murthy**

April 2026

A Note On AI Collaboration

This book was developed **in dialogue with ChatGPT**, an artificial intelligence language model developed by OpenAI.

The role of AI in this work was neither that of an authority nor an originator of ideas. It did not determine the questions asked, the positions taken, or the conclusions drawn. Those emerged from the author's long engagement with systems thinking, science, philosophy, and reflective inquiry.

AI served instead as a thinking companion—a tool for exploration, clarification, and coherence. It helped articulate questions more precisely, test the internal consistency of arguments, refine language, and maintain structural discipline across a wide and complex inquiry.

At no point was AI treated as a source of truth.

Its value lay in supporting the process of inquiry, not in supplying answers. Where uncertainty remained, it was preserved. Where boundaries between science, philosophy, and metaphor required care, those boundaries were respected.

This collaboration itself reflects one of the central themes of this book: that intelligence—whether human or artificial—is most meaningful when it supports understanding rather than substitutes for it, and when it amplifies reflection rather than authority.

All interpretations, judgements, and responsibilities for the content of this book rest solely with the author.

SECTION I

Foundations Of Space, Time, Consciousness, And Creation

Essence:

The primordial scaffolding of existence—space as stage, time as flow, consciousness as the first distinction.

Part 1

The Nature Of Space: The Invisible Architect Of Reality

Opening Question

What is space?

Is it merely an empty stage on which matter performs, or is it an active participant— quietly shaping motion, form, causality, and even existence itself?

For centuries, space was assumed to be *nothing*. Modern physics now tells us it is *everything that makes everything else possible*.

1. From Emptiness to Entity

In everyday language, space is what remains when all objects are removed. In classical physics, space was a passive container— unchanging, absolute, and indifferent to what occupied it.

This view dominated until the early 20th century.

Then came a radical realisation:

Space is not a void. It has structure. It responds. It evolves.

Space is not *where* reality happens. Space is part of *what* reality is.

2. Space Before Einstein: A Silent Background

In Newtonian physics:

Space is absolute and fixed.

Time flows uniformly for all observers.

Objects move *through* space, but space itself never reacts.

This framework worked extraordinarily well for centuries. Yet it carried a hidden assumption:

Space exists independently of matter and energy.

That assumption turned out to be incomplete.

3. Einstein's Revelation: Space Is Not Passive

Einstein's theory of General Relativity overturned the classical picture.

Key insight:

Mass and energy do not merely occupy space—they shape it.

Space and time merge into a single entity: **spacetime**.

Matter tells spacetime how to curve.

Curved spacetime tells matter how to move.

Gravity is no longer a force *in* space. It is a consequence of space itself being distorted.

A falling apple and an orbiting planet are not "pulled"—they are following the natural geometry of spacetime.

4. Space as Geometry, Not Substance

Importantly, spacetime is not a material fabric in the ordinary sense.

It has:

- No atoms
- No particles
- No substance that can be grasped

Yet it has:

- Curvature
- Geometry
- Measurable effects

This makes space a unique kind of entity:

Real, but intangible
Invisible, yet decisive

It is the silent architect that shapes every trajectory in the Universe.

5. The Dynamic Nature of Space

Space is not static.

It can:

- Expand (as in cosmic expansion)
- Stretch (as gravitational waves pass through)
- Warp near massive objects
- Change its geometry over cosmic time

The Universe is not expanding *into* space. Rather, **space itself is expanding**.

Every galaxy moves away from every other not because of motion through space, but because the space between them is growing.

6. Quantum Hints: Space Is Not Empty Even When Empty

At the smallest scales, quantum physics deepens the mystery.

So-called "empty space" contains:

Quantum fluctuations

Virtual particle – antiparticle pairs

Vacuum energy

This implies:

Even when nothing is present, space is active.

Emptiness is not absence. It is a restless field of possibility.

7. Space as a Condition for Existence

Nothing physical can exist *without* space.

Objects need spatial extension

Motion requires spatial separation

Causality depends on spatial relationships

Space is not just a backdrop—it is a **precondition** for form, interaction, and differentiation.

Without space:

No distance

No structure

No individuality

In that sense, space is the first quiet step from unity to multiplicity.

8. A Philosophical Reflection

If matter is *what* exists, and energy is *how* it acts, then space is **that which allows existence to take shape at all.**

It does not command. It does not announce itself. It simply *permits*.

Space does not speak—yet everything that exists speaks through it.

Poetic Essence

Space is not empty.
It is patient.

It does not move,
yet all motion bends to its shape.

Invisible, it governs paths.
Silent, it decides destinies.

Before matter took form,
space made room for becoming.

Closing Thought

To understand reality, one must first unlearn the idea that space is nothing.

Space is the **first architecture of existence—**

the quiet framework within which time flows, matter forms, and reality unfolds.

Only after understanding space

can, we meaningfully ask the next question:

What, then, is time?

End of Part 1

Part 2

The Nature of Time

Opening Question

What is time? Is it a river flowing from past to future, a ticking clock governing all change, or a deeper ordering principle that makes change itself possible?

We live inside time, measure everything by it, and yet struggle to say what it truly is.

1. Time as Experience Before Physics

Human intuition treats time as obvious:

The past is fixed

The present is fleeting

The future is open

We remember yesterday, experience now, and anticipate tomorrow. This psychological arrow of time feels universal and absolute.

Yet physics tells a subtler story.

2. Classical Time: A Universal Clock

In Newtonian physics:

Time flows uniformly everywhere

All observers share the same "now"

Time is independent of motion and matter

Time, like space, was considered a silent background—a cosmic clock ticking identically across the Universe.

This assumption, too, would not survive.

3. Einstein's Shock: Time Is Not Absolute

Einstein revealed that time is not universal.

Key insights:

Time passes differently for observers in motion

Time slows in strong gravitational fields

There is no single, objective "now"

Time is inseparable from space. Together, they form **spacetime**.

Two observers moving differently do not merely disagree about motion—they disagree about *time itself*.

4. Time as a Dimension, Not a Flow

In relativity:

Past, present, and future all exist within spacetime

Events are ordered, not flowing

The Universe resembles a "block" of spacetime rather than a moving moment

Nothing in the fundamental equations demands a flowing present.

This leads to a disturbing idea:

The flow of time may not be a property of the Universe, but a feature of human consciousness.

5. The Arrow of Time: Why Time Feels Directional

If physics allows time symmetry, why do we experience direction?

The answer lies in **entropy**.

Systems naturally evolve from order to disorder

The Second Law of Thermodynamics defines a direction

Memory, causality, and aging follow this arrow

We do not age because time flows.

We age because change, under constraint, is not symmetric.

The past is remembered because it had lower entropy. The future feels open because entropy has not yet increased there.

Time's arrow emerges—it is not fundamental.

6. Time at the Quantum Scale

At quantum scales, time becomes even stranger.

Fundamental equations work equally forward and backward

No preferred direction exists

Time behaves as a parameter, not an operator

Some approaches to quantum gravity suggest:

At the deepest level, **time may not exist at all.**

It may emerge only when many degrees of freedom interact—much like temperature emerges from molecular motion.

7. Time as Change, Not Container

A radical but increasingly accepted view:

Time is not something in which change happens—time *is* the ordering of change.

If nothing changes, time has no meaning.

Without processes:

No clocks

No duration

No before or after

Time is relational, not absolute.

8. The Human Experience of Time

Despite physics dissolving time into geometry and statistics, our lived experience remains undeniable.

We suffer loss. We grow. We anticipate.

This tension between physical time and experienced time is one of the deepest unresolved questions in science and philosophy.

Time may be emergent—but its consequences are real.

Poetic Essence

Time does not march.
It arranges.

Nothing moves through time;
events simply stand in order.

Memory casts the past behind us,
hope projects the future ahead.

Between them, consciousness
creates the illusion of now.

Closing Thought

Space provides the structure of existence.
Time provides its ordering.

If space answers *where*,
time answers *when*—or perhaps more profoundly, **why sequence exists at all.**

Having explored space and time separately,
we must now confront a deeper question:

What binds them together—and what lies beneath both?

End of Part 2

Part 3

When (and How) Consciousness Arose

Opening Question

When did consciousness arise? Was it a late arrival in an ancient, indifferent universe— or was it present, in some subtle form, from the very beginning?

And more provocatively:

Is consciousness a product of the Universe, or is the Universe a product of consciousness?

1. The Conventional View: Consciousness as a Late Emergence

Modern science typically holds that:

The Universe began without life

Matter organised into stars and planets

Life emerged on at least one planet

Consciousness arose much later through biological evolution

In this view, consciousness is:

Local

Fragile

Accidental

Dependent on complex brains

The Universe existed for billions of years without awareness.

But this explanation leaves critical questions unanswered.

2. The Hard Problem of Consciousness

Even if brains produce behaviour, computation, and intelligence, one question remains unresolved:

Why is there subjective experience at all?

Why does neural activity feel like something from the inside?

No physical description of neurons, signals, or algorithms fully explains:

Awareness

Qualia

The sense of "I am"

This gap is not minor. It may be fundamental.

3. Consciousness as an Emergent Phenomenon

One line of thought suggests:

Consciousness emerges when matter reaches sufficient complexity

Like temperature or pressure, it is not fundamental

It appears only at higher levels of organisation

This explains *when* consciousness appears, but not *why* subjective experience accompanies complexity.

Emergence describes correlation—not origin.

4. A Deeper Possibility: Consciousness as Fundamental

An alternative view, gaining renewed attention, proposes:

Consciousness is not produced by the Universe—it is woven into its fabric.

In this perspective:

Consciousness is not localised

Individual minds are expressions, not generators

Awareness precedes form

This does not require mystical assumptions. It arises naturally from unresolved features of physics itself.

5. The Observer in Quantum Physics

Quantum mechanics introduces a remarkable fact:

Measurement affects outcomes

Observation is not passive

Reality behaves differently when unobserved.

While interpretations differ, the role of the observer cannot be ignored.

At minimum:

Reality is not fully independent of observation

Information and measurement are fundamental

The classical separation between subject and object breaks down

This hints—cautiously—that awareness may not be an afterthought.

6. When Could Consciousness Have Arisen?

Three broad possibilities emerge:

1. **Late emergence:** Consciousness arises only with biological brains.
2. **Gradual emergence:** Proto-consciousness exists in simpler systems, intensifying with complexity.
3. **Primordial presence:** Consciousness exists as a fundamental aspect of reality, manifesting through form.

Each option carries profound implications.

The third radically reframes the question:

Consciousness did not *arise*—it *expressed itself*.

7. Consciousness and Time

If time itself is emergent, then asking *when* consciousness arose may be ill-posed.

Perhaps:

Consciousness is not located *in* time

Time unfolds *within* conscious ordering

Awareness provides the stage on which sequence appears

This view aligns with the idea that:

Time orders experience

Consciousness gives meaning to order

8. Individual Minds as Local Expressions

Under this lens:

The brain is not a generator but a receiver or filter

Individual consciousness is localised, not isolated

Death may be an end of expression, not of awareness itself

This does not negate neuroscience. It reframes its role.

The brain shapes *how* consciousness appears—not *that* it appears.

Poetic Essence

Consciousness may not have arrived late.
It may have been waiting.

Before stars ignited,
before time learned to flow,
awareness may have known itself
without form.

Brains did not invent consciousness;
they gave it a voice.

Closing Thought

If space provides structure
and time provides sequence,
then consciousness may provide **meaning.**

The Universe may not be a machine that accidentally woke up.

It may be a process of awareness
slowly learning to see itself.

With this possibility in mind,
we now turn to a deeper inquiry:

What is the relationship between consciousness, information, and the very fabric of reality?

End of Part 3

Part 4

Om, Information, And The Pulse Of Creation

Opening Question

What is the Universe made of at its deepest level? Matter? Energy? Fields? Or something more abstract—**information**?

And if information lies at the root of reality, could ancient intuitions about *sound*, *vibration*, and *Om* be early metaphors for this deeper truth?

1. Creation as Vibration, Not Construction

Across cultures, creation is rarely described as fabrication. It is described as **emanation**.

> "Let there be..."
>
> The Word
>
> The primordial sound
>
> The first vibration

Creation begins not with objects, but with **a disturbance in stillness**.

In this sense, Om is not a religious artifact. It is a symbol—pointing to **the moment structure emerges from silence**.

2. Om as a Metaphor, Not a Mechanism

Om should not be understood as an acoustic sound in space. There were no ears before creation.

Rather, Om represents:

A primordial ordering principle

A pulse that initiates differentiation

The transition from unmanifest to manifest

It is not *heard*. It is *expressed*.

Just as equations are not physical objects but shape reality, Om symbolises the **logic of becoming**.

3. Information: The Modern Language of Creation

Physics increasingly describes reality in informational terms:

Particles are excitations of fields

Fields are governed by equations

Equations encode relationships

Relationships define outcomes

In this view:

Matter is secondary

Information is primary

Laws are patterns of constraint

Reality behaves less like solid machinery and more like **executing code**.

4. "It from Bit": Reality as Encoded Order

The physicist John Wheeler proposed a striking idea:

"It from bit"—physical reality arises from informational distinctions.

A bit is the simplest choice: yes/no, on/off. Yet from vast networks of such distinctions emerge:

Space

Time

Particles

Forces

Life

Creation, then, is not assembly—it is **encoding**.

5. The Pulse: Discreteness Beneath Continuity

Modern physics reveals that reality is not infinitely smooth.

Energy comes in quanta

Time may have a smallest interval

Change occurs in discrete steps

Beneath apparent continuity lies **rhythm**.

This aligns with the ancient intuition:

Creation unfolds as pulses

Order arises through repetition

Stability emerges from oscillation

The Universe does not hum continuously. It **beats**.

6. Om and the Laws of Physics

Seen this way, Om can be interpreted as:

The first constraint

The initial rule-set

The primordial "instruction"

Not a sound in space, but the **permission for structure to arise**.

Physics asks *how* the laws operate. Om points to *that there are laws at all*.

7. Consciousness, Information, and Meaning

Information alone is not meaning.

A book unread is information without experience. Meaning arises only when information is **registered**.

This reconnects us to Part 3:

Consciousness is required for meaning

Information requires interpretation

Creation without awareness is incomplete

Thus:

Om is not blind vibration

It is **order capable of being known.**

8. The Silence Beneath the Pulse

Equally important is what precedes Om.

Before vibration:

Stillness

Symmetry

Undifferentiated potential

Om is not the beginning of existence—it is the beginning of **expression**.

Silence is not absence. It is **unmanifest fullness**.

Poetic Essence

Before form, there was stillness.
Before stillness, possibility.

Then came the pulse—
not heard, but enacted.

Not a sound in space,
but the rule that allowed space to be.

Om is not noise.
It is permission.

Closing Thought

Space gives room.
Time gives sequence.
Consciousness gives meaning.

Information gives **structure**.

Om stands at the threshold—
where silence becomes law,
potential becomes pattern,
and reality begins to speak
without uttering a word.

The next question follows naturally:

If creation is structured thought, is the Universe itself a form of mind?

End of Part 4

Part 5

The First Word: Is Language The Root Of Creation?

Opening Question

If Om represents the primordial pulse, and information structures reality, then a deeper question arises:

Is the Uinverse written? Not in letters or sounds, but in **relations, rules, and symbols**.

Is creation fundamentally an act of **language**?

1. Language Beyond Speech

When we hear "language," we think of words, grammar, and sound. But language, at its core, is more general:

A system of symbols

Rules of combination

Constraints that generate meaning

Mathematics is a language. Physics is written in it.

The Universe does not *speak*—yet it **obeys syntax**.

2. Laws of Nature as Grammar

Physical laws behave like grammatical rules:

They restrict what can happen

They define allowable combinations

They preserve consistency across scales

Particles "obey" equations. Fields "follow" symmetries.

This obedience is not voluntary—it is structural.

Reality behaves as though it understands a language without being conscious of it.

3. The Unreasonable Effectiveness of Mathematics

Mathematics, a product of human abstraction, describes nature with astonishing precision.

This raises a profound puzzle:

Did humans invent mathematics?

Or did we discover the Universe's native syntax?

The latter seems unavoidable.

Mathematics appears less like a tool and more like **a translation of reality's own grammar**.

4. The First Word as Constraint, Not Sound

The "First Word" should not be imagined as an audible utterance.

It is:

The first rule

The first distinction

The first constraint that limits infinite possibility

Without constraint:

Nothing differentiates

Nothing persists

Nothing becomes

Creation begins not with abundance, but with **restriction**.

5. Language Creates Worlds

In human experience:

Language shapes perception

Names create categories

Symbols stabilise meaning

Similarly, in physics:

Constants fix relationships

Symmetries define invariances

Conservation laws preserve identity

Language does not add substance. It adds **order**.

6. Thought, Language, and Creation

If creation is structured, and structure requires rules, then creation resembles **thought** more than construction.

Thought:

Organises without material

Operates through symbols

Generates complexity through simple rules

This does not imply a human-like mind. It implies **intelligibility** at the root of existence.

7. Is the Universe Self-Describing?

A remarkable possibility emerges:

The Universe may be both the text and the reader.

Laws describe the Universe

The Universe enacts the laws

Conscious beings arise who can read the description

This loop is not accidental. It suggests **reflexivity**—a system capable of understanding itself.

8. Language, Information, and Consciousness

Information without language is noise. Language without consciousness is inert.

Meaning arises at their intersection.

This triad—**language, information, awareness**—may be the minimal architecture of reality.

Poetic Essence

In the beginning was not sound,
but structure.

Not a word spoken,
but a rule enacted.

Creation did not shout.
It defined.

And in defining,
it allowed meaning to arise.

Closing Thought

If the Universe is intelligible,
it is because it is **structured like a statement**.

If it can be known, it is because it was **never mute**.

The First Word was not spoken in air—
it was written into possibility itself.

This leads to the next, inevitable inquiry:

If creation is structured like thought,
is the Universe itself a mind in motion?

End of Part 5

Part 6

Primordial Thought: Is The Universe The Mind Of God?

Opening Question

If creation is structured like language, if laws behave like grammar, and if information precedes matter,

then a daring question presents itself:

Is the Universe merely governed by laws—or is it the expression of a deeper, thinking order?

Is the Universe **a machine**, or is it **mind-like** at its root?

1. Thought as Order, Not Emotion

To speak of "mind" at a cosmic scale is not to imagine emotions, intentions, or preferences.

Primordial thought, if it exists, would mean:

Order without effort

Structure without construction

Intelligence without personality

It would be **pattern-generating capacity**—not cognition as humans know it.

2. The Universe as an Executing Logic

The Universe unfolds with:

Consistency

Predictability

Mathematical coherence

From simple equations emerge:

Galaxies

Chemistry

Life

Conscious observers

This resembles not randomness, but **execution of a rule-set**.

The Universe behaves less like an accident and more like a **self-unfolding logic**.

3. Laws Without a Lawgiver?

Physics describes *how* laws operate, but not *why* these laws—or why laws exist at all.

Why this fine structure? Why these constants? Why intelligibility?

To say, "they just are" ends inquiry. To ask, "what gives rise to lawfulness itself?" opens it.

Primordial thought is one possible answer—not a conclusion, but a framing.

4. God as Ground, Not Entity

In this framing, "God" is not:

A being within the Universe

A creator operating in time

An external engineer

Rather, God represents:

The ground of intelligibility

The source of order

That which allows laws to exist

This aligns with the idea of **immanence**, not intervention.

5. The Universe Thinking Itself into Form

If thought is the capacity to generate structure, then the Universe may be thought made visible.

Not deliberate thinking, but **self-consistent unfolding**.

Stars, atoms, and minds are not added later—they are natural expressions of this unfolding.

Creation is not an event. It is **ongoing articulation**.

6. Consciousness as Localised Reflection

Human consciousness may not be an anomaly.

It may be:

A localised reflection of a deeper order

A point where the Universe becomes aware of itself

A mirror, not a source

This does not elevate humans above the cosmos. It situates them *within* its logic.

7. Thought Without Time

If time itself is emergent, then primordial thought cannot be temporal.

It does not occur *before* creation. It is **that which allows before and after to exist**.

Creation unfolds in time; the ordering principle does not.

8. The Risk of Metaphor—and Its Necessity

All language here is metaphor.

Yet metaphor is unavoidable when thought approaches its own limits.

Physics gives equations. Philosophy gives questions. Metaphor builds bridges between them.

The idea of cosmic mind is not doctrine—it is **an interpretive lens**.

Poetic Essence

Not a hand that builds,
but a logic that unfolds.

Not a voice that commands,
but an order that permits.

The Universe does not think as we do—
yet thought may be how it exists at all.

Closing Thought

If the Universe is intelligible,
if it generates minds capable of understanding it,
and if its laws resemble language more than machinery,

then asking whether it is mind-like
is not mystical indulgence—
it is a legitimate philosophical response
to scientific depth.

The next step is inevitable:

If reality unfolds as thought, how does matter acquire form and mass within this order?

End of Part 6

SECTION II

Symmetry, Asymmetry, And The Architecture Of Form

Essence:

How balance enables structure and asymmetry generates form, identity, and direction.

PART 7

Chirality, The Self And The Mirror Of God

Opening Question

If the deepest laws of reality are symmetric, why does the world appear asymmetric— and why does identity arise only when symmetry breaks?

For thought to become matter, it must first acquire direction. That direction enters the Universe through asymmetry.

1. The Science: When Symmetry Is Not Enough

At the deepest levels of physical law, symmetry reigns.

The equations of physics—classical mechanics, electromagnetism, quantum field theory—are profoundly symmetric. They do not prefer left over right, past over future, matter over antimatter. In principle, the Universe could have emerged as a perfectly balanced mirror of itself.

Yet it did not.

Reality, as we observe it, is *decisively asymmetric*.

This asymmetry is not random. It is structured. Directional. Persistent.

Physics gives this asymmetry a precise name: **chirality**.

A chiral object is one that cannot be superimposed on its mirror image. Your left and right hands are the simplest example—mirror reflections, identical in form, yet fundamentally non-interchangeable. Rotate them, flip them, examine them from every angle: they will never coincide.

This phenomenon appears everywhere:

Amino acids in living systems are exclusively **left-handed**

DNA twists into a **right-handed helix**

Weak nuclear interactions violate mirror symmetry

Biological development distinguishes left from right at the cellular level

Life itself chooses one orientation and rejects the other

Chirality is not an imperfection of nature. It is a **choice embedded into reality**.

And that choice makes structure possible.

2. From Matter to Meaning: Why Asymmetry Creates Identity

A perfectly symmetric universe would be static.

No preference means no direction. No direction means no evolution. No evolution means no individuality.

Symmetry provides *possibility*. Asymmetry provides *actuality*.

Chirality breaks the deadlock.

Once a system chooses left instead of right—or right instead of left—it acquires:

Orientation

Memory

History

Path dependence

In physics, this allows particles to behave differently under transformation. In chemistry, it allows molecules to interact selectively. In biology, it allows life to emerge with specificity and function.

In consciousness, it allows **selfhood**.

The moment symmetry breaks; *identity is born*.

3. The Self as a Chiral Reflection

Now the question deepens.

If matter itself acquires identity through asymmetry, what of the *self*—the experiencing center that says, *"I am"*?

The self, too, is a chiral phenomenon.

Each conscious being is a **non-superimposable reflection** of a deeper, formless reality. We are not copies. We are orientations.

Vedanta expresses this insight with profound simplicity:

> **Brahman** is non-dual, formless, infinite.
>
> **Jiva** is Brahman reflected through limitation, context, and embodiment.

The mirror does not create the image. Nor does the image exhaust the source.

Just as left and right hands arise from the same underlying structure yet remain distinct, the individual self-arises as a **chiral reflection of the Absolute**—similar in essence, irreducible

in identity.

Multiplicity does not deny unity. It *expresses* it.

Chirality suggests something deeper than asymmetry alone. It points to a principle of expression—where the same underlying reality can manifest in distinct, non-interchangeable forms. What appears as difference may not arise from separate origins, but from how structure allows one essence to take on direction, identity, and orientation.

4. The Mirror of God

Chirality offers a striking metaphysical metaphor.

God—or ultimate reality—is not split into opposites. Rather, the infinite becomes visible only when reflected asymmetrically.

Form arises when symmetry allows itself to be broken.

In this view:

God is not diminished by creation

The many do not fragment the One

Individuality is not a fall, but a feature

Each self is a mirror—not a replica.

The mirror introduces orientation. Orientation introduces perspective. Perspective introduces experience.

And experience is the Universe knowing itself from within.

Poetic Essence

Symmetry dreams.
Asymmetry awakens.

From perfect balance,
the universe tilts—
and in that tilt,
a face appears.

Not identical,
not opposite,
but oriented.

The self is not a shard of God,
but a mirror turned at an angle,
catching infinity
in a finite frame.

Reflective Closure

Chirality reveals a quiet truth:

Symmetry creates the possibility.
Asymmetry creates the world.

Without symmetry, there is no structure.
Without asymmetry, there is no self.

The Universe does not merely exist—it *chooses*.
And in that choice,
it becomes many
without ever ceasing to be One.

End of Part 7

PART 8

The God Particle: Physics Meets The Metaphysical Field

Opening Question

If asymmetry gives the Universe orientation and identity, what gives those identities weight, substance, and resistance—allowing form to become matter?

The Scientific Turn: From Identity to Mass

Chirality explains *why forms differ*. It does not yet explain *why forms endure*.

Orientation alone is not enough.

Particles may possess identity, charge, and interaction rules— yet without mass, they would move at the speed of light, unable to bind, unable to settle, unable to build complexity.

A universe of pure symmetry-breaking but no mass would be dynamic, elegant—and uninhabitable.

Physics confronted this mystery directly.

Why do some particles have mass while others do not? Why does the Universe allow solidity instead of pure flux?

The answer emerged not as a particle first, but as a **field**.

1. Higgs Field: The Invisible Grantor of Substance

Modern physics reveals a startling fact:

Mass is not intrinsic.

Particles do not *carry* mass by default. They **acquire** it through interaction.

The Higgs field permeates all of space. It is everywhere, always present, even in apparent emptiness.

When particles move through this field:

Some interact strongly → they become massive

Some interact weakly → they remain light

Some do not interact at all → they remain massless

The Higgs boson—the so-called *God Particle*—is not the source of mass. It is merely the **brief, observable ripple** in this omnipresent field.

The field is primary. The particle is the whisper.

2. Why This Matters Beyond Physics

This discovery carries profound philosophical weight.

Mass—so fundamental to our everyday intuition—is not a thing. It is a **relationship**.

Substance arises not from isolated existence, but from **immersion in a deeper, invisible order**.

Matter gains weight because it participates.

This mirrors a much older insight.

3. The Metaphysical Parallel

In Vedantic thought, Brahman is not a being among beings. It is the **field of being itself**.

Forms do not exist independently. They gain reality by participation.

Just as:

Particles gain mass through the Higgs field

Forms gain stability through an underlying substratum

The field is not seen. Its effects are everywhere.

The particle points to the field, but never exhausts it.

Calling the Higgs boson the *God Particle* was never a theological claim— yet it inadvertently named a metaphysical truth:

What gives reality weight is not what appears, but what **pervades**.

4. From Chirality to Embodiment

Now the arc becomes clear:

Symmetry provides lawful possibility

Asymmetry (chirality) provides identity and orientation

The Higgs field provides mass, resistance, and persistence

Form becomes *matter* when identity encounters a sustaining field.

The Universe does not merely differentiate—it **settles**. And in settling, it becomes habitable.

Poetic Essence

The universe learned to lean,
then learned to stay.

A field unseen slowed the dance of light,
taught motion to hesitate, taught form to linger.

Weight was not imposed.
It was received.

And in that quiet resistance,
the cosmos learned how to hold itself.

Reflective Closure

The Higgs field invites a radical rethinking:

Reality is not built from things.
It is shaped by fields.

What appears solid
is sustained by the invisible.

Matter is not heavy because it exists—
it is heavy because it **belongs**.

And perhaps the deepest truth whispered by the God Particle is this:

Substance arises not from isolation,
but from participation in a deeper order.

The inquiry now turns again:

If mass itself depends on an unseen field,
what does this say about time, change, and impermanence?

The answer unfolds in what follows.

End of Part 8

PART 9

The Hidden Majority Of Matter

Opening Question

If matter gives structure to time and motion, what does it mean that most of the matter in the Universe is invisible to us?

1. The Scientific Discovery: Missing Matter

Astronomy once assumed a simple rule: what we see is what exists.

Stars shine. Galaxies glow. Matter reveals itself through light.

But careful measurement shattered this assumption.

Galaxies rotate too fast for their visible mass. Clusters of galaxies remain bound when gravity should tear them apart. Gravitational lensing bends light where no luminous matter exists.

The conclusion is unavoidable:

2. Most of the Universe's matter does not emit, absorb, or reflect light.

This unseen component—called **dark matter**—outweighs ordinary matter by nearly **five to one**.

The stars, planets, atoms, and bodies we know are not the main substance of the cosmos.

They are the exception.

3. What Dark Matter Does (Without Being Seen)

Dark matter does not announce itself. It does not glow, radiate, or collide in familiar ways.

Yet it **acts**.

It shapes the formation of galaxies

It anchors large-scale cosmic structures

It determines how matter clumps over cosmic time

It governs motion without revealing form

Without dark matter:

galaxies would never have formed

stars would not have stabilised

cosmic history would look radically different

The visible universe rides atop an **invisible gravitational scaffold**.

Structure precedes appearance.

4. Time, Change, and the Unseen

This has profound implications.

If time is measured by change, and change follows motion, and motion is governed by gravity—

then **time itself unfolds under the influence of something unseen**.

The tempo of cosmic evolution is set not by what shines, but by what silently holds.

Dark matter does not participate in chemistry, life, or perception—yet it determines the stage on which all of them arise.

The most influential presence in the Universe is also the least perceptible.

5. A Philosophical Shift

This discovery forces a quiet humility.

Reality is not obligated to reveal its foundations.

What is fundamental may be:

silent rather than expressive

indirect rather than experiential

structural rather than participatory

The Universe is not built from what appears, but from what **persists and constrains.**

Visibility is not a measure of importance.

6. The Metaphysical Resonance

Ancient metaphysics anticipated this inversion.

In Vedantic thought:

the gross (visible) depends on the subtle (invisible)

form arises from that which has no form

the manifest rests upon the unmanifest

Dark matter is not consciousness. But it echoes a metaphysical pattern:

What supports the world does not need to appear within it.

The unseen is not absence. It is **architecture.**

Poetic Essence

Most of the Universe
does not announce itself.

It does not shine,
does not speak,
does not arrive in color or sound.

Yet it holds the galaxies together,
paces the dance of time,
and gives the cosmos its shape.

The brightest things
are not the strongest.

Reflective Closure

Dark matter teaches a quiet lesson:

Reality is governed not by prominence,
but by presence.

What we experience
floats upon what we cannot see.

And perhaps impermanence itself—
the fleeting nature of forms—
is possible only because something unseen remains.

The visible changes.
The invisible endures.

The inquiry now turns again:

If unseen matter shapes the Universe in silence,
how does the cosmos respond
when massive structures move, collide, and change?

The answer unfolds in what follows.

End of Part 9

SECTION III

Cosmic Motion, Waves, And Extreme Reality

Essence:

The dynamic universe of curvature, rhythm, horizons, and gravitational extremes.

PART 10

When The Cosmos Trembles: Mass, Motion, And The Music Of Spacetime

Opening Question

If unseen mass silently shapes the Universe, what happens when massive structures move, collide, and change? Does spacetime merely host events—or does it respond to them?

1. The Scientific Breakthrough: Spacetime Is Not Silent

For a long time, space was imagined as a passive stage and time as a steady clock.

Events happened *within* spacetime, but spacetime itself was thought to remain unaffected— a neutral container for cosmic drama.

Einstein changed this view decisively.

In General Relativity, mass and energy do not merely occupy space. They **curve spacetime**.

And when massive objects accelerate— when black holes spiral inward, when neutron stars collide— that curvature does not remain static.

It **moves**.

The universe trembles.

These tremors propagate outward as **gravitational waves**— ripples in spacetime itself, traveling at the speed of light.

In 2015, a century after Einstein predicted them, human instruments finally heard this cosmic music.

The universe did not just glow. It **rang**.

2. What Gravitational Waves Reveal

Gravitational waves are unlike any signal detected before.

They are not vibrations *in* space. They are vibrations *of* space.

They carry no color, no charge, no medium. They stretch and compress spacetime as they pass, altering distances by less than a fraction of a proton— yet carrying colossal energy.

From them, we learn:

Spacetime is dynamic, elastic, and responsive

Motion leaves an imprint on the very fabric of reality

Cataclysmic events echo across the cosmos long after they occur

The universe remembers motion.

3. From Structure to Rhythm

Dark matter provides the hidden scaffold. Ordinary matter supplies visibility. Motion introduces **rhythm**.

With gravitational waves, reality reveals a new dimension of order— not static geometry, but **cosmic cadence**.

The universe does not merely exist. It **vibrates**.

Galaxies rotate. Stars orbit. Black holes merge.

And spacetime answers each movement with a corresponding tremor.

This is not chaos. It is choreography.

4. A Philosophical Turn: Being as Vibration

This discovery quietly reframes existence.

If spacetime itself responds to motion, then reality is not a frozen structure.

It is a **process**.

Being is not defined solely by form or mass, but by **interaction, change, and resonance**.

What is most fundamental is not stillness, but responsiveness.

Reality is relational.

5. The Metaphysical Echo

Ancient traditions spoke of the Universe as sound, rhythm, or pulse.

Not metaphorically—but structurally.

In Vedic thought, creation is not a one-time event. It is a continuous unfolding, sustained by vibration.

Gravitational waves offer a modern parallel:

The cosmos does not merely expand. It **sings**.

Each massive event adds a note. Each collision leaves a trace.

The Universe is not silent between stars. Its silence is simply beyond ordinary hearing.

Poetic Essence

The cosmos shuddered,
and spacetime listened.

A distant collision
sent a whisper across billions of years,
barely touching Earth—
yet enough to bend our rulers
and awaken our instruments.

The Universe is not still.

It breathes,
it stretches,
it remembers motion
as music.

Reflective Closure

Gravitational waves reveal a profound truth:

Reality is not built once and left alone.
It is continually **re-shaped by motion**.

Mass tells spacetime how to curve.
Motion tells spacetime how to sing.

And in these faint cosmic tremors,
we glimpse a Universe that is alive to its own becoming.

What moves, matters.
What matters, resonates.

The inquiry now deepens:

If spacetime itself vibrates and responds,
what then is time—
the measure of motion,
or something far more mysterious?

The answer unfolds in what follows.

End of Part 10

PART 11

Black Holes And The Nature Of Time

Opening Question

If time measures change and motion, what happens to time when motion approaches its ultimate limit and gravity becomes absolute?

1. The Extreme Laboratory: Black Holes

Black holes are not exotic curiosities at the edge of physics. They are **natural consequences** of gravity taken to its logical extreme.

When sufficient mass collapses into a small enough region:

spacetime curves infinitely

escape becomes impossible

light itself is trapped

At the boundary—the **event horizon**—something extraordinary occurs.

From the perspective of a distant observer:

clocks near the horizon slow dramatically

processes appear to freeze

time stretches toward infinity

From the perspective of the falling object:

nothing unusual happens at the horizon

time flows normally

crossing is uneventful

Time, it seems, does not behave universally.

2. Time Is Not Absolute

Black holes force a radical realisation:

Time is not a background parameter ticking uniformly everywhere. It is **woven into spacetime** and shaped by gravity.

Near a black hole:

time dilates

causal order distorts

the future and past blur in unfamiliar ways

At the singularity—where known physics breaks down— our equations no longer agree on what *time* even means.

This is not merely a technical failure. It is a conceptual one.

Time itself resists being defined independently of structure.

3. The Arrow of Time Under Question

Ordinarily, time feels directional.

Entropy increases. Memories accumulate. Causes precede effects.

But black holes complicate this intuition.

Information appears to disappear behind the horizon

Hawking radiation suggests slow evaporation

The fate of information challenges the idea of irreversible loss

If information is preserved, then time's arrow may not be fundamental.

What we experience as irreversible flow may be an emergent feature— not a law written into the Universe.

4. Flow to Geometry

In black hole physics, time behaves less like a river and more like a dimension that can:

stretch

twist

slow

effectively stop

Time becomes **geometrical**, not experiential.

What feels like "flow" may arise from how conscious systems move through spacetime— not from time itself moving.

The Universe may not be *in time*. Time may be **in the Universe.**

5. A Metaphysical Reflection

Ancient philosophies hinted at this long ago.

Time was seen not as ultimate, but as a mode of perception— a way consciousness orders experience.

Black holes, through physics rather than poetry, echo this insight.

They suggest that:

time is contingent

duration is contextual

eternity is not endless time, but **timelessness**

Where structure dissolves, so does the clock.

Poetic Essence

At the edge of a black hole,
time loosens its grip.

Clocks hesitate.
Moments stretch.
The future waits.

What falls inward
does not lose time—
it leaves time behind.

Perhaps time was never flowing.
Perhaps we were.

Reflective Closure

Black holes teach a humbling lesson:

Time is not the master of reality.
It is one of its expressions.

What we call "now"
is shaped by motion, gravity, and perspective.

And in the deepest gravitational wells,
time itself bows out of the story.

Where gravity becomes infinite,
time becomes optional.

The inquiry now turns again:

If time itself can bend, slow, or dissolve,
does reality end at the boundaries we observe—
or does existence extend beyond our cosmic horizon?

The answer unfolds in what follows.

End of Part 11

PART 12

Beyond The Horizon: The Multiverse And The Many Faces Of Existence

Opening Question

If time and causality dissolve at cosmic extremes, does reality end at the horizon we observe—or is our universe one expression among many?

1. The Scientific Provocation: One Universe—or Many?

Modern cosmology began with a modest assumption: the observable universe is *the* Universe.

Yet the deeper our theories reach, the more this assumption strains.

Several independent lines of inquiry point toward a startling possibility: **our universe may not be alone.**

Among them:

- **Inflationary cosmology**, suggesting space may eternally spawn bubble universes
- **Quantum mechanics**, where multiple outcomes coexist mathematically
- **String theory**, predicting vast landscapes of possible physical laws

Fine-tuning puzzles, where life-permitting constants appear improbably precise

None of these *prove* a multiverse. But together, they make it difficult to insist on cosmic singularity.

The universe may be vast not only in size, but in **number**.

2. What "Beyond the Horizon" Means

A horizon is not an edge. It is a **limit of access**.

Light from beyond a certain distance has not yet reached us. Events beyond that boundary are not absent— they are simply **unobservable**.

Inflation extends this idea dramatically.

If space expands faster than information can travel, then regions of reality become permanently causally disconnected.

Each region may carry:

different physical constants

different particle properties

different histories

Not different places in the same Universe— but **different universes altogether**.

3. The Many Faces of Existence

If a multiverse exists, then existence itself is plural.

What we call "laws of nature" may be **local bylaws**, not universal decrees.

Gravity may differ elsewhere. Time may behave differently. Even dimensionality may vary.

Our universe becomes one **face of reality**, not its entirety.

This does not diminish meaning. It reframes it.

Uniqueness no longer rests on exclusivity, but on **context**.

4. Philosophical Reorientation

The multiverse challenges a deep human instinct: that reality must be singular to be real.

Yet plurality does not negate truth.

A forest does not invalidate a tree. A library does not erase a book.

Meaning does not require being the only one. It requires being **this one**.

If countless universes exist, then identity is not cosmic centrality—it is **situated significance**.

5. Echoes from Ancient Thought

Long before modern cosmology, ancient traditions spoke of layered realities.

Cycles of creation and dissolution. Multiple realms of existence. Worlds within worlds.

Not as scientific claims, but as metaphysical intuitions:

Reality is larger than what appears. Finality is an illusion born of limited vision.

The multiverse gives these intuitions a contemporary mathematical voice.

Poetic Essence

Beyond the horizon,
the Universe may repeat itself—
or differ beyond recognition.

Other skies.
Other laws.
Other beginnings.

Yet here we are,
in this Universe,
under these constants,
asking these questions.

Meaning does not shrink
when reality expands.

It sharpens.

Reflective Closure

The multiverse, if real, delivers a quiet lesson:

Reality is not obligated to be singular
to be significant.

Our universe may be one ripple
in an ocean of possibilities—
but it is the ripple in which
we awaken, observe, and wonder.

Existence does not need to be unique to be meaningful. It needs only to be experienced.

End of Part 12

SECTION IV

Life As Code: From Cells To Self-Awareness

Essence:

Life is seen as structured information—from molecular logic to proto-awareness.

“Every animal leaves traces of what it was; man alone leaves traces of what he created.”

—Jacob Bronowski

Life leaves form, adaptation, and continuity. Human life adds a new layer: the capacity to encode intention into language, tools, institutions, and memory. In that movement, biology does not end—it becomes reflective. Civilisation is life becoming self-aware enough to leave not only traces of existence, but traces of meaning.

PART 13

The Living Code: From Cells To Self-Aware Systems

Essence

Life is not merely chemistry in motion. It is *structured information*—encoded, processed, preserved, and transformed across scales.

1. The Question Beneath Biology

When we ask, *"What is life?"* we often receive answers framed in substances:

- Carbon-based molecules
- Proteins and nucleic acids
- Metabolism and reproduction

Yet these descriptions, while accurate, are incomplete.

They tell us **what life is made of**, but not **what life is doing**.

A deeper question emerges:

Is life defined by its material components—or by the **logic that animates them**?

2. Life as Information, Not Substance

At its most fundamental level, life behaves like a **code-driven system**.

DNA is not merely a molecule; it is an **instruction set**

RNA is not just a messenger; it is an **interpreter**

Proteins are not static structures; they are **executors of logic**

What distinguishes living matter from non-living matter is not complexity alone but **organised informational flow**.

A rock contains atoms. A cell *uses* atoms—to store memory, make decisions, repair itself, and adapt.

Life, therefore, is **matter disciplined by information**.

3. The Cell: Nature's First Computational Unit

A single living cell is astonishingly sophisticated.

Within its microscopic boundaries, it performs functions analogous to:

Memory storage (genetic material)

Signal processing (chemical and electrical gradients)

Error correction (DNA repair mechanisms)

Decision-making (gene expression on/off states)

Self-maintenance (homeostasis)

Importantly, the cell does not merely react—it **regulates**.

It senses its environment, evaluates internal states, and responds selectively.

This is not consciousness. But it is **proto-agency**.

4. Code That Reads Itself

One of the most remarkable features of life is that its code:

Builds the machinery that reads the code

Repairs the code when it is damaged

Copies the code with variation

This recursive loop—*code → reader → code*—has no analogue in non-living systems.

A book does not rewrite itself. A program does not evolve unless designed to do so.

Life does both—*naturally*.

This self-referential architecture is the seed from which complexity grows.

5. From Single Cells to Collective Intelligence

As life evolves, cells do not merely multiply. They **coordinate**.

Bacteria communicate via quorum sensing

Cells specialise and form tissues

Neural networks emerge from electrical signalling

Brains arise as predictive engines

At each stage, the underlying principle remains the same:

6. Distributed information processing across adaptive networks

The leap from unicellular life to multicellular organisms is not just biological—it is **organisational**.

Life begins to act *as a system of systems*.

In South America, scientists have discovered vast underground tunnel systems—not formed by conventional geological processes, nor built by humans, but carved by extinct megafauna such as giant ground sloths.

These structures, some stretching for hundreds of meters, reveal a striking truth: Life does not wait for theory. It acts, adapts, constructs, and solves—often at scales that appear engineered—long before intelligence becomes self-aware enough to describe what it is doing.

What we call "engineering" may, in many cases, be a late human recognition of patterns that life has already mastered through evolutionary time.

7. Emergence: When the Whole Knows More Than the Parts

No single neuron is conscious. No individual cell knows the organism.

Yet consciousness emerges.

This is not magic. It is **emergence**—a phenomenon where new properties arise when information flows across sufficient complexity, integration, and feedback.

Crucially:

Emergent properties cannot be found by inspecting parts in isolation

They exist only at the **system level**

Life, therefore, is not reducible to molecules alone. Nor is mind reducible to neurons alone.

Both arise from **organised relational patterns.**

8. Is Life Substrate-Independent?

This realisation leads to a profound possibility:

If life is fundamentally informational, could it exist in other substrates?

Biology answers with carbon. But logic does not insist on carbon.

Artificial life, digital organisms, and self-learning systems raise an uncomfortable yet inevitable question:

Is DNA *the only* living code?

Or is it *one expression* of a deeper principle?

Life may not be defined by *what it is made of*, but by *how it processes, preserves, and evolves information*.

As the boundary between mind and machine begins to blur, it invites a deeper reflection: consciousness may not be an attribute of any single form, but a principle that expresses itself through different structures—biological or otherwise—when conditions allow.

9. Toward Self-Awareness

Self-awareness does not appear suddenly.

It emerges gradually as systems:

Model their environment.

Model themselves within that environment.

Predict future states.

Adjust behaviour accordingly.

From reflex to learning, from learning to memory, from memory to identity.

The living code thickens, layers upon itself, and eventually turns inward.

When a system can say—implicitly or explicitly— *"This is happening to me"* the threshold of self-awareness has been crossed.

A Bridge, not a Conclusion

This part does not claim that cells are conscious. Nor that consciousness is inevitable.

It establishes something subtler and more important:

Life is the long experiment through which information learns to organise itself.

In the next part, we will confront a sharper question:

If life is code, and awareness emerges from organisation—

Are cells already aware in some minimal sense? And why does life insist on chirality, asymmetry, and handedness from its very foundations?

End of Part 13

Part 14

The Mirror Of Life: Are Cells Aware, And Why Are They Chiral?

Opening Reflection

Life does not announce itself. It begins without witnesses, without memory, without language.

Yet from a single cell— silent, asymmetric, responsive— arise beings who wonder, who remember, and who ask why they exist at all.

Somewhere between chemistry and consciousness, life makes its first quiet choice.

Opening Questions

Are cells, in any sense, aware? If not aware as we are, do they possess a more primitive sensitivity—a proto-awareness—that precedes thought?

And why is life, at its deepest molecular level, **irreversibly chiral**—choosing one "handedness" and excluding its mirror forever?

Are these two facts—*responsiveness* and *asymmetry*—connected? Or are they independent accidents of biology?

1. The Cell: More Than a Molecular Machine

A living cell is often described as a biochemical factory. But this metaphor conceals more than it reveals.

A cell:

Maintains order against entropy

Actively regulates its internal environment

Interprets chemical signals

Responds selectively rather than randomly

Initiates repair, division, or programmed death

These are not passive reactions. They are **context-dependent responses**.

The cell does not merely undergo change—it *manages* change.

Modern biology increasingly recognises the cell as:

A self-organising system

An information-processing entity

A goal-directed structure oriented toward survival and continuity

This does not make the cell conscious. But it does make it **non-mechanical** in the classical sense.

2. Awareness as a Continuum, not a Switch

Human awareness feels unified, reflective, and self-referential. Cellular behaviour appears automatic and unconscious.

But nature rarely operates by sudden jumps.

Between the two lie gradations:

Bacteria navigating chemical gradients

Immune cells distinguishing self from non-self

Neurons responding selectively to patterns

None of these *thinks*. Yet all of them **discern**.

This suggests that awareness may not suddenly appear with brains, but may **evolve as an increasingly refined capacity to register, respond, and integrate information.**

Cells do not know. But they *sense*.

They do not reflect. But they *respond meaningfully*.

Awareness, then, may be layered— not created from nothing, but **deepened through structure**.

3. The Enigma of Chirality: Life's Irreversible Choice

Alongside responsiveness, life displays a more puzzling feature: **absolute asymmetry**.

All known life on Earth obeys a strict rule:

Amino acids are left-handed

Sugars in DNA and RNA are right-handed

Proteins fold asymmetrically

Biological interactions depend on handedness

Mirror-image chemistry does not work.

This is not a preference. It is a **commitment**.

Once chosen, life never returns to symmetry.

Why should life, emerging from symmetric physical laws, insist on asymmetry?

4. Chirality as the Condition for Function

Chirality introduces **direction.**

Without it:

Enzymes could not bind selectively

DNA replication would fail

Metabolic pathways would lose specificity

Biological memory would collapse

Perfect symmetry allows many possibilities— but **no reliable action.**

Chirality breaks equivalence. It enables orientation, selectivity, and irreversibility.

Life does not merely require complexity. It requires **non-interchangeability**.

5. From Physical Asymmetry to Biological Identity

The Universe itself is subtly asymmetric:

The weak nuclear force violates mirror symmetry

Matter dominates over antimatter

Time flows in one direction

Life inherits this asymmetry—and intensifies it.

A chiral molecule is no longer replaceable by its mirror. A living system is no longer reversible without loss.

This is the birth of **identity**.

To live is not merely to exist, but to exist in a way that **cannot be mirrored without consequence**.

6. Cells as Mirrors, Not Thinkers

Cells do not think. They do not wonder about their existence.

Yet they embody the **preconditions of selfhood**:

Sensitivity

Selectivity

Commitment

Direction

They are mirrors—not of consciousness yet—but of a Universe that favours **structured differentiation over perfect balance**.

Life does not begin with awareness. It begins with **responsiveness bound to asymmetry**.

7. Life as Pattern / Living Code

At this level, life reveals itself less as a substance and more as a **pattern**— a stable yet evolving organisation of information, responsiveness, and irreversible choice. Cells are not important merely because they are made of carbon and chemistry, but because they instantiate a living logic: sensing, selecting, committing, and preserving coherence over time.

Matter provides the medium, but **pattern provides continuity**. In this sense, life is not confined to what it is made of but defined by how it organises itself— a principle that will become crucial as we move from individual living units to collective and emergent forms of intelligence.

Poetic Interlude

A molecule turns left
when it could have turned right.

A cell responds
where silence was possible.

A path is chosen—
and the mirror is abandoned.

From this small refusal of symmetry,
memory begins.

A cell does not ask who it is.
It only turns—
left, not right.

It binds where binding matters,
breaks where breaking saves,
and remembers without memory.

Before thought, it chooses.
Before choice, it orients.
Before awareness,
it commits to a direction
the mirror cannot follow.

Closing Reflection

Life does not begin with thought.
It begins with **commitment**.

Before awareness, there is orientation.
Before meaning, there is direction.
Before the self, there is a structure that cannot be reversed without loss.

The cell stands at this quiet threshold—
not aware of itself,
yet already shaped by the same logic
that will one day make selfhood inevitable.

If life is best understood as an organised pattern of responsiveness and irreversible choice— and if individual cells exhibit only limited sensitivity— how do such patterns scale, integrate, and transcend individuality to produce collective and emergent intelligence?

And as intelligence integrates across organisms, societies, and systems, does it remain distributed and fragmented— or does it converge toward something resembling a cosmic mind?

End of Part 14

SECTION V

Intelligence, Destiny, And The Emergent Cosmos

Essence:

Intelligence as the Universe awakening to itself—purpose, emergence, and fleeting identity.

Structure over substance

Constraint over metaphor

Relations over reification

Emergence over absolutism

Dynamics over statics

Part 15

Emergent Intelligence—From Collective Minds To The Cosmic Mind

Opening Reflection

No single cell knows the body. No single ant knows the colony. No single neuron knows the mind.

And yet— bodies move with purpose, colonies solve problems, and minds awaken to themselves.

Intelligence appears, not by addition, but by **integration**.

Opening Questions

If individual living units possess only limited awareness, **how does intelligence arise that exceeds any one of them?**

How do simple responses combine into meaning, and local interactions give rise to global understanding?

Is intelligence merely multiplied when minds gather— or does something fundamentally new emerge?

And if emergence scales repeatedly, does it stop with human minds— or does it point toward a **cosmic intelligence** underlying reality itself?

1. Emergence: When the Whole Becomes More Than the Parts

Emergence occurs when:

Interacting units follow simple rules

Feedback loops amplify patterns

Global behaviour arises that no unit individually contains

Classic examples include:

Bird flocks forming coordinated motion

Ant colonies optimizing resource paths

Neural networks generating cognition

Markets displaying collective dynamics

No single element "plans" the outcome. Yet order appears.

This is not mystery—it is **non-linear organisation**.

Emergent intelligence is not located *in* any part. It exists **between** parts.

2. From Cells to Minds: Scaling Patterns of Intelligence

Life already demonstrated this principle long before brains existed.

Cells specialised. Tissues coordinated. Organs integrated.

A nervous system did not replace cells—it **connected them**.

Neurons themselves are simple:

They fire or they don't

They respond locally

But when organised into dense, recurrent networks, they produce:

Memory

Emotion

Self-models

Awareness of awareness

Mind is not stored in neurons. Mind **emerges from their relations.**

3. Collective Minds: Intelligence Beyond the Individual

The same principle repeats at higher scales.

Human intelligence does not end at the skull.

It extends into:

Language

Culture

Institutions

Science

Technology

Civilisations remember what individuals forget. Knowledge persists across generations.

No single human "knows" civilisation—yet civilisation **knows itself through us.**

This is **distributed cognition**:

Intelligence as a networked phenomenon

Meaning as a collective construction

Progress as cumulative integration

We are not isolated minds. We are **nodes in larger cognitive systems**.

4. Technology and Artificial Emergence

Modern technology has accelerated this process.

Digital networks:

Link billions of minds

Preserve collective memory

Enable rapid coordination

Artificial intelligence systems:

Learn from aggregated human knowledge

Detect patterns no individual can see

Act as amplifiers of collective cognition

Importantly, these systems do not introduce a new principle. They **extend an old one**.

Intelligence has always scaled by:

Connectivity

Feedback

Integration

Technology simply removes friction.

5. The Pattern Repeats: Toward the Cosmic Scale

At every level we observe the same arc:

Units interact

Patterns stabilise

Intelligence emerges

Integration deepens

Cells → Organisms Organisms → societies Societies → planetary systems

The question becomes unavoidable:

Is intelligence a local accident—or a universal tendency?

If the Universe consistently produces systems capable of:

Self-organisation

Information integration

Increasing awareness

Then intelligence may not be an anomaly. It may be **the Universe becoming aware of itself**.

6. The Cosmic Mind: Science Meets Vedanta

Vedantic philosophy speaks of **Brahman**— not as a thinking entity, but as an all-pervading intelligence from which individual minds arise.

Modern science does not assert this metaphysically— but it increasingly recognises that:

Information is fundamental

Fields underlie particles

Laws precede objects

The convergence is subtle but profound:

Intelligence may not *belong* to individuals. Individuals may be **expressions of a deeper intelligence**.

Poem

A thought is never alone.
It leans on language,
inherits memory,
and echoes older minds.

What thinks through me
has thought before—
in cells,
in species,
in stars.

Perhaps the mind
is not in us.

Perhaps
we are in the mind.

Closing Reflection

Emergent intelligence reshapes how we see ourselves.

We are not isolated thinkers
accidentally placed in a vast universe.

We are **participants in an unfolding cognitive process—**

local expressions of a pattern that scales, integrates, and deepens.

From the cell's responsiveness
to the mind's reflection
to the collective intelligence of civilisations,

the same logic persists:

Awareness grows by connection.

Leading Question

If intelligence consistently arises as systems integrate— from cells to minds, from minds to collectives— is this capacity for intelligence already *encoded* in the fabric of reality, or does it genuinely *emerge* through interaction and organisation?

End of Part 15

Part 16

Encoded Or Emergent?

Opening Reflection

Some things feel planned. Others simply *happen*.

A crystal grows by rules written into its lattice. A crowd forms patterns no one intended.

Between what is *pre-written* and what *arises unexpectedly*, reality seems to hesitate.

So does intelligence.

Opening Questions

Is intelligence already woven into the fabric of the Universe, waiting patiently to be uncovered?

Or does it arise only when conditions align— through interaction, feedback, and collective organisation?

In other words: **Is intelligence encoded, or does it emerge?**

And if it emerges, what does that tell us about the nature of reality itself?

1. What Do We Mean by "Encoded"?

When something is *encoded*, its outcome is implicit from the start.

Examples are familiar:

A seed contains the blueprint of the tree

A crystal grows according to fixed molecular symmetry

Physical laws constrain what is possible

In this view:

The Universe unfolds what was already latent

Complexity is revealed, not created

Intelligence is a realisation of hidden potential

Vedantic thought often resonates here: **the effect exists in the cause** (*satkāryavāda*).

Nothing truly new appears— it only becomes manifest.

2. What Do We Mean by "Emergent"?

Emergence describes a different phenomenon.

Here, **novel properties appear at the collective level** that do not exist in the individual components.

Crucially:

No single part contains the whole

No central controller directs the outcome

The pattern arises *from interaction itself*

This is not mystery. It is observable—every day.

3. Everyday Experiences of Emergence

a. The Traffic Jam

No driver wakes up intending to create a traffic jam.

Each person:

follows simple rules

responds locally

tries to reach their destination

Yet congestion **emerges**.

The jam:

is real

affects everyone

cannot be reduced to any single driver

It exists only at the **collective level**.

b. Language

No individual "invented" a language.

Languages:

evolve through use

change without permission

develop grammar no one designed

Meaning emerges from shared patterns of communication, not from isolated speakers.

c. Markets

No single buyer or seller sets the market price.

Prices emerge from:

- countless local decisions
- feedback loops
- collective expectations

The market "knows" something no participant fully understands.

These are not encoded outcomes. They are **emergent realities**.

4. Emergence in Nature and Mind

The same logic appears at deeper levels.

- Temperature emerges from molecular motion
- Life emerges from chemistry
- Mind emerges from neurons
- Intelligence emerges from networks

A neuron is not intelligent. A brain is.

Nothing in the neuron *contains* the mind— yet the mind is undeniable.

Emergence does not violate physical laws. It **transcends reduction**.

5. A Subtle Tension: Constraints vs Creativity

Emergence does not mean "anything goes."

It always occurs **within constraints**:

physical laws

boundary conditions

environmental limits

So, the real question is not:

encoded *or* emergent

But:

how much is constrained, and how much is created?

Perhaps reality provides:

a structured space of possibilities

within which novelty can arise

The Universe may not script outcomes— but it **permits creativity.**

Poem

No one plans the jam,
yet the road comes to a halt.

No neuron knows the mind,
yet thought appears.

The rules are simple.
The result is not.

Between order and surprise,

something new is born.

Closing Reflection

The encoded view emphasises **order and inevitability.**
The emergent view emphasises **novelty and becoming.**

Reality seems to require both.

Without encoding, there is chaos.
Without emergence, there is no creativity.

Intelligence may not be fully written into the Universe—
yet the Universe seems remarkably capable
of **bringing intelligence forth.**

Leading Question

If intelligence is neither fully pre-written nor purely accidental— if it emerges within constraints rather than from a fixed script—

what distinguishes the fleeting forms it takes from the deeper principles that endure?

End of Part 16

Part 17

The Fleeting And The Essential

Opening Reflection

Everything changes. And yet— not everything is lost.

Forms arise, linger briefly, and dissolve. But something persists through their passing, quietly carrying identity forward.

We live between what passes and what remains.

Opening Questions

What, in this Universe, is truly temporary—and what is fundamental?

Are lives, minds, and civilisations merely fleeting ripples, or do they express something deeper that endures?

If intelligence emerges and evolves, **what part of it fades, and what part persists?**

1. The World of the Fleeting

The fleeting is easy to see.

Cells are born and die

Organisms age and perish

Thoughts appear and vanish

Civilisations rise and fall

Nothing embodied escapes impermanence.

Even stars exhaust their fuel. Even galaxies drift and deform.

At every observable scale, **form is temporary**.

This is not failure. It is design.

2. Impermanence as a Creative Principle

Change is often mistaken for fragility. But without impermanence:

Evolution would stall

Learning would freeze

Adaptation would cease

The fleeting allows:

experimentation

diversity

renewal

What endures unchanged cannot evolve.

Impermanence is not the enemy of meaning— it is its **enabler**.

3. The Search for the Essential

Yet beneath this flux, something remains stable enough for continuity to exist at all.

Consider:

Physical laws that persist across cosmic time

Mathematical relations that never age

Patterns that reappear across scales

Even life itself, though embodied forms perish, **continues as a pattern.**

The essential is not a thing. It is a **principle of persistence.**

4. Identity Beyond Form

A river is never the same water— yet it remains the same river.

A mind changes daily— yet identity feels continuous.

What persists is not matter, but **organisation and relation.**

In modern physics, fields outlast particles. In biology, patterns outlast organisms. In culture, ideas outlast individuals.

Identity, it seems, is **carried by structure, not substance.**

5. The Fleeting as Messenger of the Essential

The fleeting is not opposed to the essential. It **reveals it.**

A wave appears and disappears, but its motion points to the ocean beneath.

Similarly:

A short-lived particle reveals a deeper field

A transient life reveals enduring principles

A fleeting mind hints at an underlying intelligence

The temporary gives form to the timeless.

Poem

A spark flares—
and is gone.

Yet fire remains.

A life passes—
yet life continues.

What fades teaches us
what does not.

Closing Reflection

We often mourn impermanence,
forgetting that without it,
nothing new could appear.

The Universe seems to preserve its essence
not by freezing forms,
but by **allowing them to pass**.

What is essential does not resist change.
It **moves through change,**
expressing itself again
and again in fleeting forms.

Leading Question

If the Universe sustains itself by allowing forms to arise and vanish—
and if life and intelligence are among these forms—

how do individual lives fit into this larger logic of interdependence, coexistence, and shared destiny?

End of Part 17

SECTION VI

Interdependence, Ecology, And The Logic Of Life

Essence:

Networks of dependence, biodiversity, parasitism, scarcity, and abundance.

Part 18

The Dance Of Dependence—On Parasitism, Biodiversity, Scarcity, And Abundance

Opening Question

Is life fundamentally a zero-sum struggle— a competition where one survives only by defeating another? Or is it a relational system, where constraint, dependence, and even conflict give rise to resilience, diversity, and abundance?

To answer this, we must step back from moral instincts and look carefully at how life actually works.

1. The Logic of Dependence

No living system exists alone.

Every organism depends on:

energy it does not create,

materials it does not own,

ecosystems it cannot control.

Plants depend on sunlight, soil microbes, water cycles, and pollinators. Animals depend on plants, prey, predators, and climate. Even apex species depend on the stability of what lies below them.

Life is not built on independence. It is built on **structured dependence**.

What appears as vulnerability at the individual level is stability at the system level.

Dependence is not a flaw in the design of life. It *is* the design.

2. Parasitism Revisited

Parasitism is often treated as a mistake—a deviation from harmony, a moral failure of nature.

Biology tells a different story.

Parasitism is:

a survival strategy,

a selective pressure,

a regulatory force.

Parasites stress hosts. Hosts evolve defenses. Ecosystems adjust.

Without stressors, systems stagnate. Without resistance, resilience never develops.

Parasitism, then, is not error— it is a **foil through which mutuality grows**.

Dependence is not weakness— it is the **grammar of life**. Disease, too, belongs within this line.

What we call disease is often not an external anomaly, but a manifestation of interaction—between organisms, environments, and internal limits.

It arises from the same conditions that enable adaptation, survival, and diversity.

What appears as breakdown at one level may reflect constraint, imbalance, or competition at another.

The same underlying logic operates across scales—from microscopic agents to complex human systems; what changes is not the structure, but how it is experienced, interpreted, and given meaning.

3. Biodiversity and the Harmony of Contrasts

Healthy ecosystems are not optimised; they are **redundant.**

They contain:

> overlapping roles,
>
> competing strategies,
>
> inefficiencies that absorb shocks.

Monocultures collapse easily. Diverse systems endure.

Predators and prey, hosts and parasites, cooperators and competitors do not cancel each other out— they stabilise the whole through tension.

Diversity is not noise. It is the **harmony of contrasts.**

Life survives not by eliminating difference, but by **weaving difference into balance.**

4. Scarcity and Abundance—A False Dichotomy

Scarcity dominates individual experience. Abundance reveals itself only at the system level.

A single tree competes for nutrients. A forest creates its own climate.

Water may be scarce to a root, but abundant to a watershed.

Constraints do not prevent life from flourishing. They *shape* how flourishing occurs.

Evolution thrives not despite limits, but because of them.

Scarcity is often local, temporary, and informational. Abundance emerges through connection, circulation, and feedback.

Life is neither endlessly generous nor brutally stingy. It is **strategic**.

5. Money as a Civilisational Layer of Reality

In human civilisation, scarcity and abundance are not only ecological conditions; they are mediated through symbolic systems, the most powerful of which is money.

To examine reality only at the physical and biological scales is incomplete. Human civilisation introduces symbolic structures that reorganise material outcomes and redefine access to them. Among these, money stands out as the primary medium through which human intention is translated into structured claims on external reality.

Technological systems can, at times, alter the cost structure of production so significantly that certain forms of scarcity begin to recede. Intelligence, once constrained by human effort, may increasingly be produced, replicated, and distributed at near-zero marginal cost.

However, such shifts do not dissolve constraint itself. They instead relocate it.

When emerging forms of abundance interact with institutions, incentives, and coordination mechanisms designed for scarcity, a period of structural tension often follows.

The resulting instability is not a failure of the system, but a signal of transition— from one regime of organisation to another.

New equilibria do not arise automatically from abundance alone; they emerge through adaptation in governance, coordination, and collective judgement.

In this sense, money does not merely measure value; it participates in the organisation of reality at the civilisational scale.

At the level of physics, matter and energy interact through fields and forces. At the level of life, organisms adapt through survival, reproduction, and selection. At the level of civilisation, intelligent beings organise effort, exchange, and power through symbols. Money is the central symbol in that architecture.

It is not merely a medium of exchange. It is a stored claim on external reality.

From Survival to Symbolic Accumulation

For most of evolutionary history, effort was tied to immediate survival: food, shelter, kin protection, and territorial security. Human societies gradually learned to store value outside the body—first in grain, livestock, and land, later in metal, currency, credit, and financial claims.

This shift altered the logic of existence. What was once direct effort became mediated effort. What was once local exchange became abstract exchange. What was once survival expanded into accumulation.

Money made this possible by enabling value to be measured, transferred, and concentrated across distance and time.

Money as an Invisible Structure

Money has no intrinsic biological utility. It cannot be eaten, breathed, or inhabited. Its power arises from a shared social agreement: that it represents command over goods, services, labour, and future effort.

In this sense, money functions as an invisible structure within civilisation. It shapes incentives without physical contact, directs behaviour without explicit force, and enables coordination across individuals who share neither proximity nor familiarity.

The visible world of markets, institutions, and status is often the surface expression of this underlying structure.

Accumulation, Asymmetry, and Extraction

Once symbolic accumulation becomes possible, asymmetry expands.

Some accumulation arises through innovation, discipline, and productive effort. Some arises through appropriation, coercive transfer, rent-seeking, or institutionalised capture—forms of extraction that may be legal, illegal, or socially normalised depending on the context.

This is not a moral judgement, but a structural observation.

When value can be represented symbolically, control over symbols can exceed control over physical production. As a result, civilisations generate both accelerated coordination and persistent inequality.

Money amplifies both creation and concentration.

Growth Is Recent; Extraction Is Enduring

The modern doctrine of continuous economic growth is historically recent. For most of human history, societies did not operate with an expectation of sustained per-capita expansion. Gains in output were often offset by population pressures, ecological limits, or systemic disruptions.

In this older condition, the dominant logic was survival and relative position, not aggregate growth.

Only in the last few centuries—with industrialisation, access to concentrated energy sources, technological scaling, and modern financial systems—did the expectation of continuous growth become central to civilisational organisation.

Human beings have always pursued resources. The expectation that all societies must continuously expand their material output is a comparatively recent development.

Closing Reflection

If scarcity shapes individual experience and abundance emerges at the system level, then money is one of the principal mechanisms through which this tension is expressed in human civilisation.

It does not create resources, but it reorganises access to them. It does not eliminate scarcity, but redistributes its perception, often transforming what is local and temporary into what appears systemic. It does not define value, but it encodes and transmits it across time and scale.

Money is not the purpose of civilisation. It is an instrument through which effort is stored, trust is mediated, and power is negotiated.

Like all such instruments, it extends human capacity while introducing new forms of asymmetry. Its significance lies not in what it is, but in how it reshapes the relationship between intention and material reality.

6. Vedanta, Ecology, and Interbeing

Beyond biological interdependence and civilisational mediation lies a deeper insight.

From a deeper lens, the Upanishads describe the world as one consciousness appearing as many.

Multiplicity is not falsehood. It is **expression**.

In this view:

> the parasite and the host are both Brahman,
>
> the virus and the vaccine participate in the same unfolding,
>
> conflict and care arise within a single field of being.

Just as a pixel in a hologram reflects the entire image, each role—however uncomfortable—may reflect a facet of cosmic intelligence.

Interdependence is not merely ecological. It is ontological.

Poem—The Dance

Life does not stand alone.
It leans.

It leans on soil and sun,
on breath and decay,
on hands unseen and forces unnamed.

One feeds.
One takes.
One resists.
One yields.

Roots borrow from stone.
Predators borrow from prey.
Even loss borrows meaning from continuity.

What seems broken
often holds the pattern together.

What seems selfish
keeps the cycle turning.

The dance is not kind,
yet it is not cruel.

It is rhythmic.
It is relational.
It is alive.

Final Reflection

Life is a dance—
and every dancer depends on others.

Some lead.
Some follow.
Some heal.
Some hurt.

But the dance goes on—
choreographed not by dominance,
but by rhythm, feedback, and interbeing.

In this sense, the Universe is not a battlefield of competing egos,
but a grand ecology—
where even tension has purpose,
and even dependence has dignity.

Life may be fleeting.
But its meaning is not measured by duration.

What matters is:

> how deeply it touched,
>
> how clearly it resonated,
>
> and how consciously it unfolded.

End of Part 18

Questions concerning depth, understanding, and the limits of inquiry are explored further in Appendix G.

Closing Synthesis

The Architecture of Reality

After traversing space and time, fields and forces, life and consciousness, intelligence and interdependence, one insight quietly binds the entire inquiry:

Reality is not built from things. It is built from relationships.

Every structure we encountered—physical, biological, cognitive, or social—arises not in isolation, but through interaction. What persists is not substance alone, but **pattern**, not permanence, but **coherence**.

Symmetry and Asymmetry: The Deep Grammar

At the foundation of reality lies symmetry—the source of stability, conservation, and law. Symmetry allows the Universe to be intelligible, repeatable, and reliable. Without it, there would be no science, no prediction, no order.

Yet symmetry alone cannot account for a Universe that evolves, differentiates, and gives rise to novelty.

That role belongs to asymmetry.

Matter survived annihilation because symmetry broke.
Life emerged because molecules acquired handedness.
Cells differentiated, brains organised, consciousness arose—all

through departures from perfect balance.

Thus, the Universe reveals a profound logic:

Symmetry creates the possibility;
asymmetry creates the world.

Reality is neither symmetric nor asymmetric in isolation.
It is the **tension and dialogue between the two** that generates form, identity, and meaning.

Identity, Life, and Intelligence

Particles appear identical because of symmetry. They acquire mass and structure through broken symmetry.

Life preserves internal order through repeating patterns, and evolves through variation and deviation.

Consciousness recognises patterns, and assigns meaning through distinction.

Intelligence—human, collective, or artificial—does not arise from control alone, but from coordination, not from hierarchy, but from interaction. Purpose itself is not always encoded in advance. Often, it **emerges through relationship**.

In this sense, intelligence is not an object to be located, but a process that unfolds when conditions align.

The Human Mirror

The same architecture that governs the cosmos appears in human life.

We seek stability yet grow through change. We long for unity, yet live through multiplicity. We desire certainty, yet insight often arrives through disruption.

A life that is too rigid cannot grow. A life that is too chaotic cannot endure.

Meaning arises, again, from balance.

Returning to the Inquiry

This synthesis does not offer a final answer to the question of reality. Instead, it clarifies the **direction of understanding**.

Reality is not a finished structure waiting to be described. It is a living architecture that reveals itself through inquiry, relationship, and participation.

We do not observe the Universe from outside it. We are shaped by the very asymmetries that make observation—and life—possible.

What endures is not certainty, but coherence. Not possession of truth, but fidelity to inquiry.

Constraints as the Source of Coherence

At its deepest level, reality appears neither arbitrary nor fully knowable. It unfolds within quiet constraints that both enable and limit what can exist.

The Universe dances to the rhythm of Planck's constant, where continuity gives way to quantised possibility; it is veiled by Heisenberg's uncertainty, which reminds us that knowledge itself has irreducible limits; and it is structured by Pauli's exclusion, which enforces differentiation, identity, and form.

Together, these principles reveal a cosmos where order does not arise despite limitation, but because of it—where structure, individuality, and coherence emerge not from freedom alone, but from the disciplined boundaries woven into existence itself.

Passing the Inquiry Forward

What emerges from this inquiry is not a finished description of reality, but a coherent architecture of understanding. Across scales and domains, reality appears less as a collection of isolated things and more as a web of relationships—structured by symmetry, enlivened by asymmetry, and shaped through emergence. The journey traced here does not conclude with answers, but with orientation: a way of seeing how diverse phenomena may belong to a single, intelligible whole.

Truth as Orientation

If this book leaves anything unresolved, it is because reality itself resists final resolution. Truth, as encountered here, is not a conclusion to be reached but an orientation to be sustained—approached through inquiry, coherence, and humility.

What we know remains provisional; what we question remains alive. To inquire honestly is already to participate meaningfully. Beyond this point, the task no longer belongs to the author, but to each reader—and to the questions they are willing to carry forward.

This synthesis completes the journey of the book, but not the inquiry itself. That task now passes—quietly—to the reader.

Appendix A—Technical Notes

Appendix A.1—Higgs Boson And Bosonic Classification

In modern physics, particles are broadly classified into two fundamental categories: **fermions** and **bosons**. This distinction is not merely taxonomic; it reflects deep principles governing how matter and forces behave in the Universe.

Fermions are particles with half-integer spin (such as electrons and quarks). They obey the Pauli Exclusion Principle, which prevents identical fermions from occupying the same quantum state. This principle underlies the structure of atoms, the stability of matter, and the richness of chemical and biological complexity.

Bosons, by contrast, have integer spin and follow Bose – Einstein statistics. Multiple bosons can occupy the same quantum state, enabling collective phenomena such as lasers, superconductivity, and coherent force transmission. Force-carrying particles—such as photons (electromagnetism) and gluons (strong interaction)—belong to this class.

The **Higgs boson** occupies a unique place among bosons. It is a **spin-0 boson**, unlike force-carrying bosons, and does not mediate an interaction in the conventional sense. Instead, it represents a **localised excitation of the Higgs field**, an omnipresent quantum field that permeates all of space.

The Higgs field plays a foundational role in the Standard Model of particle physics by enabling particles to acquire **mass**. Elementary particles interacting with this field experience resistance to acceleration, which manifests as inertial mass. Importantly, it is the *field*—not the transient Higgs boson itself—that performs this role. The Higgs boson is observable only fleetingly, appearing momentarily in high-energy particle collisions before decaying.

This distinction highlights a deeper principle: **what is most fundamental is often not what is most visible**. The Higgs boson serves as an experimental signature, confirming the existence of a deeper, invisible structure—the Higgs field—much as a ripple reveals the presence of water beneath the surface.

Within the context of this inquiry, the Higgs mechanism illustrates how **enduring structure can underlie fleeting form**. Identity and consistency in the physical world arise not from permanent particles, but from stable fields and relationships. This insight resonates with broader themes explored throughout the book: that reality is better understood as a network of underlying processes and constraints than as a collection of isolated, enduring objects.

Appendix A.2—Magnetic Constant μ_0 (A historical note on units and meaning)

In classical electromagnetism, the magnetic constant, denoted by μ_0 (mu-zero), appears as a fundamental parameter governing magnetic interactions in vacuum. For much of the twentieth century, its value was written as:

$\mu_0 = 4\pi \times 10^{-7}\ \text{N}\cdot\text{A}^{-2}$

At first glance, this exact numerical value can appear puzzling, especially when compared with other physical constants that are measured experimentally rather than defined. The reason lies not in deep physics, but in historical choices of units.

In the original formulation of the International System of Units (SI), μ_0 was defined exactly, not measured. This definition was chosen to simplify electromagnetic equations and to align electrical and mechanical units in a coherent framework. As a consequence, the value of μ_0 appeared "large" or artificial compared with other constants of nature.

Its electric counterpart, the permittivity of free space (ε_0), was then derived indirectly through the relation:

$c^2 = 1 / (\mu_0\, \varepsilon_0)$

where *c* is the speed of light. In this framework, μ_0 and ε_0 were bookkeeping devices within a human-constructed measurement system, rather than independently revealed properties of nature.

A significant shift occurred with the **2019 revision of the SI system**, in which several fundamental constants—including the elementary charge (*e*) and Planck's constant (*h*)—were fixed by definition. As a result, μ_0 is no longer an exact defined constant, but a **measured quantity**, determined experimentally like other physical parameters.

This transition reflects a deeper philosophical insight: **constants are not merely numbers, but agreements between theory, measurement, and convention**. Some constants encode genuine physical regularities; others reflect the coordinate systems and units chosen by observers.

Within the broader inquiry of this book, μ_0 serves as a reminder that even in the most precise sciences, **description and convention intertwine**. The laws of nature may be invariant, but the numerical forms in which we express them are shaped by human choices. Recognising this distinction helps prevent confusion between **physical reality itself** and the **languages we use to describe it**.

Appendix A.3—Dark Matter: Observational Puzzles And Inferred Reality

Modern cosmology and astrophysics reveal a striking mismatch between what is **observed** and what is **expected** based on visible matter alone. Galaxies rotate too rapidly at their outer edges, galaxy clusters remain gravitationally bound despite insufficient luminous mass, and large-scale cosmic structures form in ways that cannot be explained by ordinary matter under known physical laws.

To account for these discrepancies, physicists infer the existence of **dark matter**—a form of matter that does not emit, absorb, or reflect light, yet exerts gravitational influence. Its presence is deduced indirectly through its effects on motion, structure, and gravitational lensing, rather than through direct detection.

Importantly, dark matter is not defined by *what it is*, but by **what it does**. It behaves as gravitational mass, shaping galaxies and cosmic filaments, while remaining otherwise elusive. Multiple candidate explanations exist, ranging from hypothetical particles (such as weakly interacting massive particles) to more exotic possibilities, though none have yet been conclusively confirmed.

An alternative line of inquiry questions whether dark matter reflects an incomplete understanding of gravity itself, particularly at galactic and cosmological scales. Modified gravity theories attempt to explain observations without invoking unseen matter, though these approaches face their own challenges and remain subjects of ongoing debate.

From an epistemic perspective, dark matter exemplifies a recurring theme in scientific discovery: **invisible structure inferred from**

consistent effects. Just as earlier generations inferred atoms before observing them directly, contemporary science confronts a reality that resists direct perception yet manifests through coherent patterns.

Within the context of this inquiry, dark matter highlights an essential insight: **what governs reality need not be directly observable**. Much of the Universe's architecture may consist of enduring, unseen components that provide stability and form to what is visible. This reinforces the broader lesson that prominence in experience is not a reliable measure of foundational importance.

Dark matter thus stands not only as a scientific puzzle, but as a conceptual reminder that reality often reveals itself **through absence as much as through presence**.

Appendix A.4—Why Time May Not Exist (As a fundamental entity)

In everyday experience, time appears self-evident: clocks tick, events unfold, and cause precedes effect. Yet in modern physics, the status of time is far less secure. Several foundational theories suggest that **time may not be a fundamental feature of reality**, but an emergent or derived concept.

In **classical mechanics**, time is treated as an external parameter—absolute, universal, and independent of physical processes. This view was decisively altered by **relativity**, where time becomes inseparable from space and dependent on the observer's motion and gravitational environment. There is no single, universal "now"; different observers experience different temporal orderings of events.

The challenge deepens in **quantum gravity**, where attempts to unify quantum mechanics with general relativity encounter what is often called the *problem of time*. In some formulations, the fundamental equations describing the Universe contain **no explicit time variable at all**. Instead, they describe relationships between physical states, with change expressed relationally rather than temporally.

From this perspective, what we call "time" may arise from:

correlations between physical systems,

the ordering of events relative to one another,

or the way memory, entropy, and causality structure experience.

In such views, time is not something the Universe *contains*, but something observers *infer*.

Thermodynamics offers a complementary insight. The familiar **arrow of time**—the distinction between past and future—is closely linked to increasing entropy. Yet entropy itself is a statistical concept, emerging from large numbers of interacting components. This suggests that the directionality of time may be a macroscopic consequence of complexity, not a microscopic rule.

Philosophically, this reframing dissolves a common assumption: that time must flow for change to occur. Instead, **change may be primary**, with time serving as a conceptual ordering imposed upon it. The Universe may consist of configurations and transitions, while "time" is the narrative we construct to describe them.

Within the broader inquiry of this book, the possibility that time is not fundamental reinforces a recurring theme: **what feels most immediate to experience may be least fundamental to structure**. Just as solidity emerges from empty space and mass arises from fields, time itself may emerge from deeper, timeless relations.

This does not render time illusory in practice. Rather, it suggests that time is **real as an experience and a description**, even if it is not a primitive building block of reality. Recognising this distinction allows us to hold both scientific rigour and experiential meaning without contradiction.

Appendix A.5—Fields Of Modern Physics

Modern physics is best understood not as a study of isolated particles, but as a layered framework of **fields**, each governing behaviour at particular scales and domains. These fields do not operate independently; rather, they overlap, constrain, and inform one another, forming a structured hierarchy of description.

At the most familiar scale, **classical fields** such as gravitational and electromagnetic fields describe macroscopic phenomena. Gravity governs the motion of planets and galaxies, while electromagnetism explains light, electricity, magnetism, and much of everyday matter interaction. These fields are continuous and deterministic in their classical formulations.

At smaller scales, **quantum fields** replace classical notions of particles. In quantum field theory, particles are understood as localised excitations of underlying fields that permeate all of space. The electron, photon, and quarks are not fundamental objects moving through a vacuum, but dynamic manifestations of deeper field structures.

The **Higgs field** occupies a special position among quantum fields. Unlike force fields, it provides a universal background that enables particles to acquire mass. Its presence illustrates how a pervasive, invisible field can shape the most basic properties of matter without being directly observable.

Beyond these lie fields associated with **cosmology and spacetime itself**. General relativity describes gravity not as a force, but as the curvature of spacetime—a geometric field responding to mass and energy. In this view, spacetime is not a passive stage, but an active participant in physical processes.

Attempts to unify these frameworks point toward still-deeper structures, often described in terms of **quantum gravity, information**, or **emergent spacetime**. While no complete theory yet exists, these efforts suggest that familiar fields may themselves arise from more fundamental relational or informational substrates.

Within the context of this inquiry, the landscape of physical fields serves as a conceptual map rather than a final inventory. Each field offers a level of description appropriate to certain questions, but none alone provides a complete account of reality. Together, they illustrate a central insight of modern science: **what we call reality is revealed through layered, interacting frameworks, not a single, privileged description.**

Appendix A.6—Information-First Reality (Vedral's perspective)

In recent decades, a growing body of thought in physics suggests that **information may be more fundamental than matter or energy**. Among the prominent voices advancing this view is physicist Vlatko Vedral, who argues that physical reality can be understood as emerging from informational relationships rather than from material substances alone.

In the traditional view, information is something *about* physical systems—a description encoded by observers. In the information-first perspective, this order is reversed: **information is primary**, and physical properties arise as expressions of how information is structured, transformed, and conserved.

This idea finds support in several domains of modern physics. In **quantum mechanics**, the state of a system is described not by definite properties, but by a wavefunction that encodes probabilities and correlations—informational content rather than physical attributes in the classical sense. Measurement outcomes depend on how information is accessed, not merely on what exists independently.

In **thermodynamics**, the link between information and entropy has long been recognised. Entropy can be interpreted as a measure of missing information about the microscopic configuration of a system. This connection deepens the suggestion that physical laws may constrain information flows as much as material processes.

Quantum information theory strengthens this view further. Phenomena such as entanglement reveal correlations that cannot be reduced to local physical properties of individual particles. Instead, the fundamental reality appears to reside in **relationships encoded as information**, rather than in separable objects.

Vedral's proposal does not deny the reality of matter or energy. Rather, it reframes them as **manifestations of informational structure**, much as temperature emerges from molecular motion or pressure from particle collisions. In this sense, information is not abstract or subjective, but an objective feature of how the Universe is organised.

Within the broader inquiry of this book, the information-first perspective offers a unifying lens. It bridges quantum physics, thermodynamics, cosmology, and even questions of life and consciousness, without collapsing them into a single explanatory claim. Information becomes a common language across domains—not a final answer, but a foundational principle through which diverse phenomena can be related.

This approach reinforces a recurring theme: **what appears solid and primary may emerge from patterns of relation**, while what seems abstract may quietly govern the structure of reality itself.

Appendix A.7—The Quantum – Classical Boundary As An Information Gradient

One of the enduring questions in physics concerns the boundary between the **quantum** and **classical** worlds. Quantum systems exhibit superposition, entanglement, and probabilistic behaviour, while classical systems appear definite, localised, and deterministic. Rather than representing two separate realms, these behaviours may be understood as arising from a **continuous gradient of information accessibility**.

At the quantum scale, systems are described by wavefunctions that encode probabilities and correlations. The full informational content of such systems is typically inaccessible to direct observation. Measurement reveals only partial information, collapsing a range of possibilities into a single outcome. Importantly, this collapse reflects not a sudden physical transformation, but a **change in how information is registered and constrained.**

As systems grow larger and interact with their environments, information becomes increasingly dispersed. Through processes such as **decoherence**, quantum correlations are rapidly entangled with countless environmental degrees of freedom. The system's quantum information becomes effectively inaccessible, giving rise to stable, classical-like behaviour. What emerges is not a new physical law, but a new **regime of information loss and averaging**.

From this perspective, the quantum – classical boundary is not a sharp dividing line. It is better understood as an **information gradient**, where the degree of quantum behaviour depends on how much coherent information can be preserved and accessed.

Small, isolated systems retain rich quantum structure; large, open systems display classical regularity.

This view helps reconcile why quantum mechanics governs all matter at a fundamental level, yet classical physics remains extraordinarily effective at human scales. Classical reality emerges when informational detail is washed out by complexity, interaction, and scale.

Within the broader inquiry of this book, the information-gradient perspective reinforces a central theme: **what we experience as reality depends on the level at which information is organised and available**. Apparent discontinuities between quantum and classical descriptions reflect limits of access, not contradictions in nature itself.

Seen this way, the boundary between quantum and classical worlds is not a wall, but a transition—one shaped by information flow, environmental interaction, and the practical limits of observation.

Appendix A.8—The Universe As A Quantum Computer (A conceptual metaphor)

The idea that the universe may be understood as a kind of **quantum computer** has emerged from developments in quantum information theory and theoretical physics. This view does not claim that the universe is literally a machine executing code but uses computation as a **conceptual metaphor** to illuminate how physical processes transform information.

In this perspective, the fundamental laws of physics act as **rules of information processing**. Quantum states evolve according to well-defined mathematical transformations, much like operations in a computation. Interactions between particles can be seen as information exchanges, while measurement corresponds to the extraction of specific outcomes from a range of possibilities.

Quantum mechanics naturally lends itself to this interpretation. Superposition allows systems to encode multiple potential states simultaneously, and entanglement links information across spatially separated systems. These features resemble the core resources of quantum computation, where information is distributed nonlocally and processed collectively rather than sequentially.

Physicists such as Seth Lloyd have explored this analogy by estimating how much information the universe could, in principle, process given its energy, size, and age. While such estimates are speculative, they highlight an important insight: **physical evolution and information transformation are deeply intertwined**.

Crucially, this metaphor must be handled with care. The universe does not compute *for a purpose*, nor does it store information *for an external observer*. Computation here is not intentional or designed; it is simply a way of describing how physical states evolve according to lawful constraints.

Within the context of this inquiry, the "quantum computer" metaphor reinforces a recurring theme: reality may be less about static objects and more about **processes that transform relational structure**. Information is not processed *by* the universe in the way a machine operates, but *as* the universe unfolds.

Seen in this light, computation becomes a bridge concept—connecting physics, information, and emergence—without reducing the universe to a device or implying that meaning or purpose is encoded in its operation.

Appendix A.9—The Transactional Interpretation: Offer Waves, Confirmation Waves, And The Handshake Of Reality

The **Transactional Interpretation (TI)** of quantum mechanics, originally proposed by John G. Cramer, offers a time-symmetric way of understanding quantum processes. Rather than treating quantum events as unfolding solely forward in time, TI proposes that physical interactions involve a **two-way exchange across time**.

In this interpretation, a quantum event begins with an **offer wave**—a standard quantum wavefunction that propagates forward in time from a source. Potential absorbers respond by sending **confirmation waves** backward in time. When an offer and a confirmation successfully match, a completed transaction forms. This mutual exchange constitutes a realised quantum event.

The resulting "handshake" is not an exchange of signals in the everyday sense, but a **consistent linking of possibilities** that satisfies conservation laws and boundary conditions across time. The transaction is global and self-consistent, selecting one outcome from many allowed possibilities.

One of the strengths of the Transactional Interpretation is that it provides an intuitive account of **quantum nonlocality** without invoking faster-than-light communication. Correlations between entangled particles are explained not by signals traveling through space, but by transactions that extend across spacetime as a whole.

Equally important is the interpretation's **time symmetry**. The fundamental equations of quantum mechanics allow solutions that move both forward and backward in time. TI takes this symmetry seriously, suggesting that temporal direction emerges from completed interactions rather than being imposed at the outset.

Within the broader inquiry of this book, the Transactional Interpretation serves as an illustration of how **causality, time, and interaction may be relational rather than sequential**. Reality appears less like a chain of events unfolding moment by moment, and more like a coherent network of constraints that settle into consistency.

Whether or not the Transactional Interpretation proves to be the definitive account of quantum behaviour, it highlights an important lesson: **our intuitions about time and causation may be too narrow for the quantum domain**. Alternative interpretations can expand understanding without altering empirical predictions, reminding us that interpretation itself is part of how reality is rendered intelligible.

Appendix A.10—Chaos, Randomness, And Hidden Order

At first encounter, many natural systems appear chaotic or random. Weather patterns, turbulent fluids, biological evolution, market fluctuations, and even quantum measurement outcomes, often resist precise prediction. Yet modern science has revealed that **apparent disorder frequently arises from underlying order**, governed by deterministic rules or statistical structure.

Chaos refers to behaviour in deterministic systems that is highly sensitive to initial conditions. Even when the governing equations are simple and fully known, small differences in starting states can lead to dramatically different outcomes over time. This sensitivity makes long-term prediction impractical, giving rise to behaviour that appears random despite being lawfully generated.

Randomness, by contrast, is often associated with unpredictability rooted in incomplete information or fundamental probabilistic limits. In classical contexts, randomness may reflect ignorance of microscopic details. In quantum mechanics, randomness appears intrinsic to measurement outcomes, though the underlying wavefunction evolves deterministically according to well-defined equations.

Crucially, chaos and randomness do not imply the absence of structure. Chaotic systems often display **hidden order** in the form of attractors, scaling laws, and statistical regularities. Similarly, quantum probabilities obey precise mathematical rules, even when individual events cannot be predicted.

The study of chaos has shown that complexity can emerge from simplicity without external guidance. Order need not be imposed from above; it can arise naturally from interactions among components governed by local rules. This insight has reshaped understanding across physics, biology, and complex systems science.

Within the broader inquiry of this book, chaos and randomness illustrate a recurring theme: **lack of predictability is not equivalent to lack of meaning or structure**. What appears disordered at one level may be coherent at another. The universe may be lawful without being predictable, structured without being rigid.

Recognising this distinction helps dissolve a common misconception—that uncertainty or variability signals fundamental disorder. Instead, uncertainty often marks the boundary between levels of description, where deeper regularities give rise to rich and sometimes unpredictable behaviour.

Seen in this light, chaos and randomness are not failures of understanding, but **signatures of complexity**, revealing how order and unpredictability coexist within the same underlying reality.

Appendix A.11—Neutrinos: Ghost Particles And The Question Of Why Anything Exists

Among the most elusive constituents of the Universe are **neutrinos**—elementary particles that interact so weakly with matter that billions pass through every square centimetre of Earth each second without notice. They carry no electric charge, possess extremely small mass, and rarely interact except via gravity and the weak nuclear force.

Neutrinos are produced in vast numbers by nuclear reactions in stars, supernovae, particle accelerators, and even within Earth itself. Despite their abundance, they are extraordinarily difficult to detect, earning the description "ghost particles."

For many years, neutrinos were assumed to be **massless**. However, experiments demonstrating neutrino oscillations—where neutrinos change identity as they travel—revealed that they must possess a small but nonzero mass. This discovery had profound implications, requiring extensions to the Standard Model of particle physics.

Neutrinos play a subtle yet crucial role in **cosmic evolution**. In the early Universe, their sheer number influenced how matter clustered under gravity. In stellar explosions, neutrinos carry away enormous energy, shaping the formation of heavy elements essential for planets and life. Though individually inconsequential, collectively they exert significant influence on cosmic structure.

From a deeper perspective, neutrinos illustrate a recurring theme in modern physics: **the most influential components of reality may**

be the least perceptible. They contribute to the Universe's energy balance and evolution while remaining almost entirely invisible to direct experience.

Within the inquiry of this book, neutrinos serve as a reminder that existence is not defined solely by what is stable, massive, or readily observable. Instead, reality appears to be sustained by a complex interplay of fleeting, subtle, and often hidden participants. The fact that such nearly undetectable entities help shape the cosmos invites reflection on a broader question: **why the Universe contains anything at all rather than nothing.**

Neutrinos do not answer this question directly. Yet their quiet presence reinforces the insight that the Universe is structured not only by dominant forces and particles, but also by marginal, elusive contributors whose cumulative effects enable complexity, continuity, and emergence.

Appendix A.12—Fermat's Principle: Extremal Paths And The Logic Of Nature

Fermat's Principle states that light traveling between two points follows a path for which the **optical travel time is extremal**—most commonly minimal, but more precisely *stationary* with respect to small variations of the path. This deceptively simple idea has profound implications across physics.

In everyday terms, Fermat's Principle explains why light bends when passing from air into water, why lenses focus images, and why mirages appear. Rather than light "choosing" a path in a conscious sense, the principle captures a deep regularity: **the realised path is the one that satisfies a global constraint**, not merely local rules.

Mathematically, Fermat's Principle belongs to a broader class of **variational principles**, where physical behaviour is derived from extremal conditions. Similar formulations appear in classical mechanics (the principle of least action), general relativity (geodesics in curved spacetime), and quantum mechanics (path integrals). Across these domains, nature appears to favour outcomes that render certain quantities stationary rather than arbitrary.

Importantly, the extremal path need not always be the shortest or fastest in the intuitive sense. What matters is not optimisation in a human sense, but **consistency under variation**. The realised configuration is the one that remains stable against infinitesimal alternatives.

From a conceptual standpoint, Fermat's Principle suggests that physical processes are not merely sequential reactions to local causes. Instead, they reflect **global coherence**, where the entire path or configuration participates in determining the outcome. The behaviour of light seems to "know" the boundary conditions in advance—not through foresight, but through lawful structure.

Within the broader inquiry of this book, Fermat's Principle reinforces a recurring insight: **nature often operates through constraints and consistency rather than step-by-step instruction**. Reality appears shaped by conditions that select viable patterns from among many possibilities, rather than by mechanical trial-and-error.

Seen this way, Fermat's Principle is not just a rule about light. It is a window into how order arises—quietly, efficiently, and without intention—through the interplay of possibility and constraint that underlies the physical world.

Appendix A.13—A Speculative Law Of Spacetime: Order From Constraint

From time to time, physics encounters proposals that attempt not to extend existing theories incrementally, but to **reframe the foundations themselves**. One such recent speculative approach suggests that spacetime, rather than being a primitive backdrop, may arise from deeper organising constraints that govern how information, energy, and causality relate.

At the heart of this idea is the proposal that **spacetime geometry is not fundamental**, but emergent—shaped by underlying consistency conditions that determine which configurations of reality can stably exist. In this view, spacetime is less a container in which events occur and more a *result* of lawful relational structure.

Such proposals often draw inspiration from multiple domains:

• Quantum information theory, where geometry appears linked to entanglement • Thermodynamics, where macroscopic order emerges from statistical constraints • Gravity, where spacetime curvature reflects energy distribution

What distinguishes this line of thought is its emphasis on **constraint over substance**. Instead of asking *what spacetime is made of*, it asks *what rules must be satisfied for spacetime to be coherent at all*. Geometry, dimensionality, and causal order then emerge as secondary features of a deeper logical framework.

It is important to note that these ideas remain **speculative**. They are not yet supported by decisive experimental evidence, nor do they replace established theories such as general relativity or quantum field theory. Rather, they represent exploratory attempts to address unresolved tensions between quantum mechanics and gravity.

Within the context of this book, the value of such proposals lies not in their immediate correctness, but in the **shift of perspective they encourage.** They suggest that the Universe may be governed less by elemental building blocks and more by conditions of consistency—by what is *allowed* rather than by what is *constructed*.

This aligns with a broader theme recurring throughout the inquiry: that reality often appears to organise itself through **global coherence, relational structure, and emergent order**, rather than through direct assembly from fundamental parts.

Whether or not this particular speculative framework endures, it illustrates a crucial lesson at the frontier of physics: when existing concepts reach their limits, progress may come not from adding complexity, but from **rethinking what we take to be fundamental.**

Appendix A.14—Time Dilation: Observer-Dependent Time

Time dilation is one of the most experimentally verified predictions of relativity. It describes how the **rate at which time passes depends on the observer's state of motion or gravitational environment**. Crucially, it does not assert that time itself slows or speeds up universally, but that **different observers measure different elapsed times between the same events**.

1. Relativistic Origin

In **Special Relativity**, time dilation arises from the constancy of the speed of light. Because all inertial observers measure light to travel at the same speed, clocks moving relative to one another must tick at different rates to preserve this invariance. A clock in motion relative to an observer is measured to run slower than one at rest in that observer's frame.

In **General Relativity**, gravitational time dilation occurs because gravity curves spacetime. Clocks deeper in a gravitational field (closer to massive objects) run slower than clocks farther away. This effect has been confirmed using precision atomic clocks placed at different altitudes and is routinely corrected for in satellite navigation systems such as GPS.

2. What Time Dilation Demonstrates (and What It Does Not)

Time dilation demonstrates that:

Time is **not absolute**

Temporal measurements are **frame-dependent**

There is no single, universal clock governing the Universe

However, time dilation does **not** explain:

What time fundamentally *is*

Why time flows

Whether time exists as a primary constituent of reality

It is therefore an **operational phenomenon**, not an ontological explanation.

3. Conceptual Significance for This Inquiry

Within the context of this book, time dilation serves as **empirical evidence** for a deeper insight developed across several chapters: namely, that time may be **relational and emergent**, rather than a fixed background parameter.

Time dilation shows that:

Different observers can each be correct, yet disagree about elapsed time

Temporal order and duration depend on physical relationships, not absolutes

Measurement precedes interpretation—clocks respond to structure, not essence

Thus, time dilation supports—but does not resolve—the central question of whether time is fundamental or emergent.

4. Placement Within the Appendix

Time dilation is placed here to:

Preserve the philosophical continuity of the main chapters

Avoid reducing inquiry into calculation

Provide technical grounding for readers seeking empirical confirmation

It stands as a **bridge between theory and experience**, illustrating how deep structural features of spacetime manifest in measurable reality—without claiming to define reality itself.

Closing Note

Time dilation reminds us that **time behaves** but does not tell us **what time is.** In that distinction lies the boundary between physics as measurement and inquiry as understanding.

Appendix A.15—Simultaneity, Relation, and the Absence of a Universal "Now"

One of the most profound implications of modern physics is the realisation that there is **no universal present moment** shared across the Universe. Events that appear simultaneous to one observer may not be simultaneous to another. This is not a limitation of perception or measurement, but a fundamental feature of how reality is structured.

In **special relativity**, simultaneity depends on the observer's state of motion. Two events judged to occur at the same time in one frame of reference may occur at different times in another moving frame. There is no privileged frame from which a global "now" can be defined.

This insight undermines a deeply ingrained intuition: that the Universe evolves through a single, synchronised sequence of moments. Instead, physics reveals a world composed of **intersecting timelines**, each associated with a particular path through spacetime.

The concept of a **worldline** captures this idea precisely. A worldline represents the history of an object or observer as a continuous path through spacetime. Time experienced by an observer is not something that flows independently, but something **accumulated along that path**. Different worldlines accumulate different amounts of time between the same events.

From this perspective, temporal order is not absolute but **relational**. What exists are events and the causal relationships between them.

"Past," "present," and "future" are not universal layers of reality, but ways of organising experience relative to a given observer.

This view does not deny the reality of time as lived and measured. Rather, it clarifies that **temporal experience is local**, shaped by motion, gravity, and interaction. The absence of a universal "now" does not render time meaningless; it renders it contextual.

Within the broader inquiry of this book, the relativity of simultaneity completes a conceptual arc begun with time dilation and extended through the problem of time itself. Together, these insights suggest that **time is not a background against which reality unfolds**, but a relational measure emerging from the structure of spacetime and the paths taken within it.

Seen this way, reality does not advance uniformly through a single present moment. It consists instead of a vast network of interrelated processes—coherent, lawful, and connected—**without the need for a cosmic clock to keep them in step**.

Appendix B—Key Concepts In Quantum Physics

B.1 Wavefunction

B.2 Bloch Sphere

B.3 Superposition

B.4 Entanglement

B.5 Probability Amplitudes

B.6 Quantum Interference

B.7 Measurement and Collapse

B.8 Quantum Decoherence

B.9 Uncertainty Principle

B.10 Quantum Fields vs Particles

Formally, quantum states are represented within an abstract mathematical structure known as Hilbert space—a linear vector space that encodes superposition and probability amplitudes.

The present appendix focuses on the physical and interpretive implications of quantum behaviour, rather than its full mathematical formalism.

Appendix B.1—Wavefunction

In quantum physics, the **wavefunction** is the central mathematical object used to describe the state of a quantum system. Unlike classical descriptions that specify definite properties such as position or velocity, the wavefunction encodes **all that can be known** about a system prior to measurement.

Crucially, the wavefunction does **not** describe a physical wave in space like a water wave or sound wave. Instead, it represents a **field of possibilities**—a structured catalogue of how a system may behave when observed.

The values of the wavefunction are generally complex numbers. While these values themselves are not directly observable, their **squared magnitude** determines the **probability** of finding a particular outcome when a measurement is made. In this sense, the wavefunction connects physical reality to statistical prediction.

Before observation, a quantum system is said to exist in a **superposition** of possible states, all encoded within the wavefunction. Measurement does not reveal a pre-existing definite value; rather, it selects one outcome from among the possibilities described by the wavefunction.

Interpretations differ on whether the wavefunction represents:

an objective feature of reality,

a state of knowledge,

or a relational description dependent on the observer and measurement context.

What is unambiguous, however, is its role: the wavefunction is the **most complete description available** of a quantum system prior to observation—a bridge between mathematical structure, probability, and physical experience.

Appendix B.2—Bloch Sphere

The **Bloch Sphere** is a geometric way to visualise the state of a simple quantum system known as a **two-level system**, or *qubit*. While the underlying mathematics is abstract, the Bloch Sphere provides an intuitive picture of how quantum states relate to one another.

The sphere represents **all possible pure states** of a qubit as points on its surface. Two opposite points correspond to the system's two definite classical-like states, often labelled $|0\rangle$ and $|1\rangle$. Any other point on the surface represents a **superposition** of these states, with different proportions and relative phases.

Movements on the Bloch Sphere correspond to **quantum transformations**:

Rotations represent controlled changes to the state (quantum gates).

Different paths encode different physical evolutions or operations.

The latitude and longitude of a point capture both probability balance and phase information.

Importantly, the Bloch Sphere shows that quantum states are not merely "either – or" but form a **continuous space of possibilities**. Two states that look similar probabilistically may still differ in phase—a difference that can have real physical consequences.

The Bloch Sphere does **not** describe spatial motion. It is a **state-space visualisation**, illustrating how quantum information is structured rather than where particles are located.

In this way, the Bloch Sphere serves as a bridge between abstract quantum mathematics and human intuition, revealing how quantum states evolve, interfere, and transform—even when they defy classical imagery.

Appendix B.3—Superposition

Superposition is a defining feature of quantum physics. It refers to the ability of a quantum system to exist in multiple possible states **simultaneously**, prior to measurement. Unlike classical systems, which occupy one definite state at a time, quantum systems encode several potential outcomes at once.

In practical terms, superposition means that a quantum system does not possess a single, determinate value for a property (such as position or energy) until it is observed. Instead, all allowed values coexist as part of the system's quantum description, with each possibility assigned a specific probability.

Superposition should not be imagined as a system being physically split or duplicated. Rather, it represents a **structured set of possibilities** encoded in the wavefunction. These possibilities can interfere with one another, producing effects—such as enhanced or suppressed outcomes—that have no classical counterpart.

A key insight is that superposition is not mere uncertainty due to ignorance. It is a **fundamental feature of quantum behaviour**, confirmed experimentally through interference phenomena. The system genuinely behaves as though multiple alternatives are present and interacting.

When a measurement is performed, the superposition gives way to a single observed outcome. The transition from multiple possibilities to one realised result raises deep interpretive questions, addressed differently across quantum interpretations. What remains consistent is that **superposition governs behaviour prior to observation**, while measurement reveals a specific manifestation.

Within the broader inquiry of this book, superposition illustrates how reality at its most fundamental level is **possibility-rich rather than determinate**. Definiteness emerges not as a starting point, but as a consequence of interaction, observation, and constraint.

Appendix B.4—Entanglement

Entanglement is one of the most distinctive and counterintuitive features of quantum physics. It occurs when two or more quantum systems become so deeply correlated that their properties can no longer be described independently, even when the systems are separated by large distances.

In an entangled state, the wavefunction describes the systems **as a whole**, not as individual parts. Measuring one system instantaneously constrains the possible outcomes of the other, regardless of the spatial separation between them. This correlation exceeds anything allowed by classical physics.

Importantly, entanglement does **not** involve the transmission of signals or information faster than light. While outcomes are correlated, they cannot be used to send controllable messages. The effect reflects a shared quantum structure, not causal influence across space.

Entanglement has been confirmed experimentally through violations of Bell's inequalities, demonstrating that no local, classical explanation can account for observed correlations. These results imply that quantum reality is **nonlocal in structure**, even though it remains consistent with relativistic limits on causation.

From a conceptual standpoint, entanglement challenges the classical idea that systems possess independent, intrinsic properties. Instead, it suggests that **relationships may be more fundamental than objects**. Properties emerge only in the context of interaction and measurement.

Within the broader inquiry of this book, entanglement exemplifies a recurring insight: reality is not fully decomposable into separate parts. At its foundation, the Universe appears woven together by correlations that transcend spatial separation, reinforcing the view that **connection precedes individuality**.

Appendix B.5—Probability Amplitudes

In quantum physics, outcomes are not described directly by probabilities, but by **probability amplitudes**. A probability amplitude is a mathematical quantity—typically a complex number—whose squared magnitude gives the probability of a particular measurement outcome.

This distinction is crucial. In classical situations, probabilities simply add. In quantum systems, **amplitudes combine first**, and probabilities are calculated only after this combination. As a result, quantum outcomes can display patterns that have no classical analogue.

Probability amplitudes can:

> **reinforce** one another, increasing the likelihood of an outcome, or
>
> **cancel** one another, suppressing outcomes that would otherwise be expected.

This behaviour underlies **quantum interference**, where the structure of possibilities—not just their likelihood—determines what is observed. Two paths that are each individually allowed may together produce a diminished or even zero probability, depending on how their amplitudes combine.

The use of complex numbers is not incidental. The phase of a probability amplitude—encoded in its complex nature—plays a decisive role in how amplitudes interact. Two outcomes with the same probability may behave very differently depending on their relative phase.

From a conceptual standpoint, probability amplitudes shift the meaning of chance in quantum theory. Randomness does not arise from ignorance alone, but from the **structured combination of possibilities** governed by precise mathematical rules. What appears unpredictable at the level of outcomes is highly constrained at the level of amplitudes.

Within the broader inquiry of this book, probability amplitudes reinforce a recurring insight: **reality is shaped not only by what can happen, but by how possibilities relate to one another**. The Universe does not merely select outcomes; it organises potentiality according to coherent, though often non-intuitive, principles.

Appendix B.6—Quantum Interference

Quantum interference is the defining feature that distinguishes quantum behaviour from classical probability.

In classical thinking, possibilities are treated independently. Probabilities are assigned to alternative outcomes and then **added directly**. The order does not matter, and unseen possibilities play no role once an outcome is chosen.

Quantum theory works differently.

In the quantum domain, what combine first are **probability amplitudes**, not probabilities. These amplitudes are complex-valued quantities that encode both magnitude and phase—a structured potential for outcomes, not outcomes themselves.

Only **after amplitudes combine** are probabilities obtained, by taking the square of the resulting amplitude's magnitude. At that point, the total probability of all possible outcomes sums to one.

This single shift—**amplitudes combine first, probabilities appear later**—explains several otherwise puzzling features of quantum reality:

- Why interference patterns arise, even when particles are sent one at a time
- Why "possible paths" influence outcomes even when they are not observed
- Why quantum behaviour is structured rather than random
- Why quantum outcomes cannot be explained as hidden classical probabilities

In quantum interference, possibilities do not merely coexist—they **interact**.

Paths reinforce or cancel one another depending on their relative phases. Some outcomes become more likely, others less likely, not because of chance, but because of structured combination.

This is why quantum behaviour is not classical probability.

Interference reveals that the quantum world is governed by **relations among possibilities**, not by isolated alternatives. Reality, at its most fundamental level, is shaped by how potential outcomes combine before any single outcome is realised.

Appendix B.7—Measurement And Collapse

In quantum physics, **measurement** is not a passive act of observation, but a process that plays an essential role in determining what is observed. Unlike classical systems, which possess definite properties regardless of observation, quantum systems are described by a range of possible outcomes prior to measurement.

Before measurement, a quantum system is represented by a wavefunction encoding multiple possibilities through probability amplitudes. When a measurement is performed, one specific outcome is realised. This transition—from a set of possibilities to a single, definite result—is commonly referred to as **wavefunction collapse**.

The term "collapse" should not be taken literally. It does not imply a physical wave suddenly imploding in space. Rather, it signifies a **change in description**: once information is obtained through interaction with a measuring apparatus, the wavefunction is updated to reflect the outcome that has occurred.

Different interpretations of quantum mechanics account for collapse in different ways. Some regard it as a physical process; others treat it as an effective description arising from interaction with the environment or the measuring device. Despite these differences, all interpretations agree on the practical fact that **measurement produces a definite result**.

A key feature of quantum measurement is its **irreversibility**. Once an outcome is registered, the prior superposition of possibilities cannot be reconstructed. This irreversibility links measurement to thermodynamics and information theory, as information about alternative possibilities is effectively lost.

Within the broader inquiry of this book, measurement and collapse illustrate a central insight: **observation participates in the formation of reality rather than merely revealing it**. The act of measurement marks the boundary where structured possibility gives way to concrete experience, highlighting the relational nature of physical description.

Seen in this way, collapse is not an anomaly to be explained away, but a reminder that at the deepest level, reality becomes definite through interaction, constraint, and the acquisition of information.

Appendix B.8—Quantum Decoherence

Quantum decoherence describes the process by which quantum systems lose their ability to exhibit coherent superposition and interference when they interact with their environment. It provides a crucial bridge between quantum behaviour and the emergence of classical reality.

In an idealised setting, a quantum system can remain isolated, preserving delicate superpositions of states. In practice, however, all systems interact—however weakly—with surrounding particles, radiation, or fields. These interactions cause the system's quantum information to become **entangled with the environment**, dispersing coherence across many degrees of freedom.

As a result, interference effects that depend on precise phase relationships become effectively unobservable. The system appears to settle into a definite state, even though the underlying quantum evolution remains lawful and continuous. Decoherence does not destroy superposition; it makes it **inaccessible.**

A key insight of decoherence theory is that it explains why quantum effects are rarely observed at macroscopic scales. Large systems interact with their environments so rapidly and extensively that coherence is lost almost instantaneously. What emerges is behaviour that closely resembles classical physics.

Importantly, decoherence does **not** by itself explain why a particular outcome occurs in a measurement. It accounts for the disappearance of observable interference, but not for the selection of one specific result among many possibilities. In this sense, decoherence complements—but does not replace—interpretations involving measurement and collapse.

Within the broader inquiry of this book, decoherence reinforces a recurring theme: **classical reality emerges through interaction, not by abandoning quantum rules**. Definiteness arises not from a fundamental change in laws, but from the redistribution and loss of accessible information.

Seen this way, decoherence reveals how complexity and environment shape what can be experienced. The boundary between quantum and classical worlds is not fixed, but dynamic defined by how coherence is preserved or lost within a web of interactions.

Appendix B.9—Uncertainty Principle

The **Uncertainty Principle**, formulated by Werner Heisenberg, expresses a fundamental limit on how precisely certain pairs of physical properties can be known simultaneously. Most famously, it states that the position and momentum of a particle cannot both be specified with arbitrary accuracy at the same time.

This limitation is **not** a result of imperfect instruments or experimental disturbance. Rather, it reflects a basic feature of quantum systems themselves. The properties in question are linked through the wave-like structure of the quantum state, making precise joint definition impossible.

At the heart of the Uncertainty Principle lies the use of **probability amplitudes** and wavefunctions. A state sharply localised in position must involve a wide range of momenta, while a state with well-defined momentum must be spread out in space. The uncertainty arises from this mathematical and physical structure, not from observer ignorance.

The principle generalises beyond position and momentum. Other pairs of quantities—such as energy and time, or different components of angular momentum—also exhibit intrinsic uncertainty relations. Together, these constraints define the limits within which quantum description operates.

Conceptually, the Uncertainty Principle challenges the classical idea that physical systems possess simultaneously well-defined properties independent of observation. Instead, it suggests that **nature constrains what can be meaningfully specified**, not merely what can be measured.

Within the broader inquiry of this book, the Uncertainty Principle illustrates a recurring theme: **limits are not deficiencies, but structural features of reality**. Precision itself is bounded by coherence, scale, and relational structure. These bounds do not undermine understanding; they shape it.

Seen this way, uncertainty is not a statement about chaos or ignorance, but a reflection of how quantum reality is organised—placing principled limits on what can be known, while enabling rich and consistent patterns of behaviour to emerge.

Appendix B.10—Quantum Fields Vs Particles

In everyday language, physics is often described as the study of particles—tiny objects moving through space and interacting with one another. Modern quantum theory, however, presents a deeper picture: **fields are fundamental, and particles are secondary manifestations**.

In **quantum field theory**, every type of particle is associated with an underlying field that permeates all of space. What we observe as a particle is a **localised excitation** or disturbance of its corresponding field. An electron, for example, is not a permanent object traveling through emptiness, but a temporary expression of the electron field.

This perspective resolves several conceptual difficulties. It explains how identical particles can be truly indistinguishable, since they are excitations of the same field rather than separate entities. It also accounts naturally for particle creation and annihilation, phenomena that have no clear analogue in classical particle-based thinking.

Fields provide the continuity and structure; particles provide the **events**. Fields exist everywhere, even in regions devoid of particles, while particles appear only when energy is exchanged or interactions occur. In this sense, particles are **contextual**, while fields are **enduring**.

The field-based view also aligns with earlier concepts introduced in this appendix. Superposition, interference, entanglement, and uncertainty arise not from particles behaving oddly, but from the wave-like, relational nature of quantum fields themselves. Measurement reveals particle-like outcomes, but the underlying dynamics remain field-based.

Within the broader inquiry of this book, the distinction between fields and particles reinforces a central insight: **what appears solid and discrete often emerges from what is continuous and relational**. Reality is not fundamentally composed of isolated things, but of structured processes that give rise to observable events.

Seen this way, particles are best understood not as the building blocks of reality, but as **signatures of deeper field dynamics**—fleeting expressions of an underlying order that persists even when no particles are present.

Appendix C—Glossary Of Key Concepts

Appendix C.1—Time

Time is commonly experienced as the flow of moments from past to future, but in modern physics it is not a universal or absolute entity.

In **classical physics**, time was treated as a uniform background—ticking identically for all observers, independent of motion or gravity. This intuition aligns with everyday experience but fails at high speeds or strong gravitational fields.

In **relativity**, time becomes **observer dependent**. Different observers, moving relative to one another or experiencing different gravitational conditions, can legitimately disagree on durations and simultaneity. There is no single, global "now" shared across the Universe.

In **quantum and foundational physics**, time occupies an even more subtle role. Some formulations treat time as an external parameter, while others suggest that time may be **emergent**—-arising from correlations, information flow, or relational change rather than existing as a fundamental ingredient.

Within this book, time is approached not as a thing that flows, but as a **relational structure**: a way of ordering events, tracking change, and expressing causal relationships.

Time, therefore, is best understood not as an absolute background of reality, but as a **context-dependent feature of how reality is described and experienced**.

Appendix C.2—Space

Space is commonly understood as the three-dimensional arena in which objects exist and events occur. In everyday experience, it appears continuous, stable, and independent of the matter within it.

In **classical physics**, space was treated as an absolute container—an unchanging stage against which motion unfolds. Objects move *through* space, but space itself remains unaffected.

With **relativity**, this view changes fundamentally. Space is no longer separate from time, nor is it passive. Instead, it forms a unified structure with time—**spacetime**—whose geometry is influenced by mass and energy. Gravity, in this framework, is not a force acting within space but a manifestation of space's curvature.

In **quantum physics**, space becomes even less intuitive. At very small scales, classical notions of smooth continuity may break down. Some theories suggest that space may have a discrete, granular, or emergent character, arising from deeper informational or relational structures.

Throughout this inquiry, space is treated not merely as an empty container, but as an **active participant in physical reality**—one that shapes motion, constrains interaction, and encodes structure.

Rather than asking where things are *in* space, modern physics increasingly asks how spatial relationships arise at all.

Appendix C.3—Spacetime

Spacetime is the unified framework in which space and time are not separate entities, but interwoven dimensions of a single structure. Events are defined not only by *where* they occur, but also by *when* they occur, forming points within spacetime rather than within space alone.

In this view, motion is no longer described as objects moving through space over time; instead, objects trace **worldlines** through spacetime. Different observers, depending on their state of motion, may disagree on distances and durations, yet all descriptions remain consistent within the same spacetime structure.

A key implication of spacetime is that **time is not universal**. There is no single, global "now" shared across all observers. What appears simultaneous to one observer may not be so to another. This challenges intuitive notions of absolute time while preserving physical consistency.

Spacetime is also **dynamic.** Mass and energy influence its geometry, bending spacetime and thereby shaping motion and causality. In dynamic systems, changes in mass – energy distribution propagate as gravitational waves—disturbances in spacetime curvature itself. In this sense, gravity is not an external force acting within spacetime, but a manifestation of spacetime's structure itself.

Within this inquiry, spacetime is treated as a **relational and geometric framework**, not merely a backdrop. It organises causality, constrains interaction, and provides the stage on which emergence unfolds—while itself remaining open to deeper explanation at quantum scales.

Appendix C.4—Field

A **field** is a physical entity that assigns a value to every point in space—or spacetime—representing the potential for interaction. Unlike particles, which appear localised, fields are **continuous** and pervasive. What we perceive as particles are often best understood as **localised excitations of underlying fields**.

In classical physics, familiar examples include gravitational and electromagnetic fields, which describe how masses and charges influence one another across distance. In modern physics, fields are more fundamental than particles: every known particle corresponds to a specific quantum field.

Fields do not merely transmit forces; they define **what can happen** at each point. They shape dynamics, enable interactions, and encode constraints. Even in what is commonly called "empty space," fields persist, fluctuating and carrying structure.

Importantly, a field is not a substance in the ordinary sense. It is a **relational construct**—a framework describing how properties and influences are distributed. Fields exist whether or not particles are present to excite them.

Within this book, fields are treated as **foundational elements of reality**, bridging structure and interaction. They exemplify a recurring theme of the inquiry: that reality is organised not primarily by objects, but by underlying relational patterns.

Appendix C.5—Mass

Mass is a measure of an entity's resistance to change in motion and its participation in gravitational interaction. In classical physics, mass is treated as an intrinsic property of matter—something an object simply *has*.

Modern physics reveals a deeper picture. At the most fundamental level, mass is not merely a substance-like attribute but an **emergent property** arising from interaction. Elementary particles acquire mass through their coupling with the Higgs field; without this interaction, many particles would remain massless and move at the speed of light.

Mass also shapes spacetime. In general relativity, mass and energy influence the curvature of spacetime, guiding how objects move. Thus, mass is both a source of inertia and a contributor to gravitational structure.

Crucially, mass is not conserved in isolation. It can be transformed into energy and vice versa, as captured by the relation between mass and energy. This underscores that mass is part of a broader relational framework rather than an independent essence.

Within the context of this inquiry, mass exemplifies a central theme: **properties we once considered fundamental often arise from deeper fields, interactions, and symmetries**. Mass is not the starting point of reality, but a consequence of how reality is structured.

Appendix C.6—Chirality

Chirality refers to a property of objects or systems that are **mirror-asymmetric**—they cannot be superimposed on their mirror image, even though the mirror image appears structurally similar.

A familiar example is the human hand: the left and right hands are mirror images, yet no rotation or translation can make them identical. This distinction is not cosmetic; it reflects a deep structural asymmetry.

In **physics**, chirality appears at multiple levels:

> In **particle physics**, certain fundamental interactions—most notably the **weak nuclear force**—distinguish between left-handed and right-handed particles. Nature, at its most fundamental level, does not treat mirror images as equivalent.
>
> In **quantum field theory**, fermions (such as electrons and neutrinos) exhibit chiral behaviour that influences how they interact and evolve.
>
> In **cosmology**, large-scale asymmetries (such as matter – antimatter imbalance) hint that mirror symmetry may be broken not only locally but cosmically.

In **chemistry and biology**, chirality is central to life:

> Biological molecules (amino acids, sugars, DNA helices) exhibit consistent handedness.
>
> Life does not randomly choose chirality; it **selects and preserves asymmetry**, enabling structure, function, and evolution.

Conceptually, chirality illustrates a recurring theme of this inquiry:

Symmetry defines possibility; asymmetry creates actuality.

Chirality shows that reality is not constructed solely from balanced, reversible laws, but from **selective asymmetries** that introduce direction, identity, and differentiation. Without chirality, many physical processes would remain neutral; with it, complexity becomes possible.

Chirality is therefore not an anomaly or imperfection. It is a **structural feature through which reality expresses form, identity, and evolution.**

Appendix C.7—Symmetry

Symmetry refers to an invariance under transformation—a property of a system that remains unchanged when viewed from different perspectives or subjected to certain operations, such as rotation, translation, or reflection.

In **physics**, symmetry is not merely aesthetic. It is foundational:

> The laws of physics appear the same regardless of **location** (spatial symmetry),
>
> regardless of **moment** (temporal symmetry),
>
> and often regardless of **orientation** or **reference frame**.

These symmetries are deeply connected to conservation principles. For example:

> Symmetry in time corresponds to conservation of energy,
>
> Symmetry in space corresponds to conservation of momentum.

Thus, symmetry acts as a **structural constraint** on what can occur. It defines the stability, predictability, and coherence of physical law.

In **quantum physics**, symmetry governs:

> Particle classifications,
>
> Interaction rules,
>
> And the mathematical structure of quantum fields.

Yet symmetry does not imply uniformity or stasis. Rather, it provides a **framework within which variation can occur**. Symmetry specifies what must remain invariant even as systems evolve.

In a broader conceptual sense, symmetry represents **unity beneath diversity.** It is the hidden regularity that allows disparate phenomena to be understood as expressions of the same underlying principles.

Within this inquiry, symmetry functions as:

The condition of intelligibility—the reason reality can be described, modelled, and partially understood at all.

However, symmetry alone does not generate structure. A perfectly symmetric universe would be featureless. Symmetry **permits** form but does not **select** it.

For that, symmetry must be complemented by its counterpart—asymmetry.

Appendix C.8—Asymmetry

Asymmetry refers to the breaking or absence of invariance—a condition in which a system changes under transformation and thus distinguishes one state, direction, or configuration from another.

While symmetry establishes what remains unchanged, **asymmetry is what allows difference, structure, and direction to arise**.

In **physics**, asymmetry is not an imperfection but a necessity:

The Universe contains more matter than antimatter,

Time appears to flow in one direction,

Biological molecules exhibit consistent handedness,

Forces do not act identically under all transformations.

These asymmetries are responsible for the **existence of form, history, and complexity**.

A well-known example is the **arrow of time**. Although many fundamental laws are symmetric with respect to time, macroscopic processes—such as entropy increase—exhibit irreversible behaviour. This asymmetry gives rise to memory, causation, and experience.

In **quantum physics**, asymmetry appears in:

Measurement outcomes,

Symmetry-breaking during phase transitions,

And the selection of specific states from a range of possibilities.

Asymmetry does not contradict symmetry; rather, it **operates within symmetric laws**. The rules may be invariant, but their realised outcomes are not.

Conceptually, asymmetry is what enables:

Distinction over sameness,

Direction over neutrality,

Identity over indifference.

Within this inquiry, asymmetry represents:

The emergence of the particular from the possible.

Without asymmetry, reality would be uniform and unstructured. Without symmetry, it would be incoherent and unintelligible. Reality requires **both**—symmetry to define what is allowed, and asymmetry to determine what actually occurs.

Together, they form the minimal tension from which structure, evolution, and meaning arise.

Appendix C.9—Quantum

The term **quantum** refers to the foundational framework of physics that describes reality at the smallest observable scales, where classical intuitions about objects, continuity, and determinism no longer apply.

In the quantum domain, physical systems are described not by definite properties, but by **states of possibility**. Outcomes are governed by probability amplitudes, and behaviour is inherently non-deterministic until interaction or measurement occurs. Concepts such as superposition, interference, entanglement, and uncertainty arise naturally within this framework.

A defining feature of the quantum description is **discreteness**. Certain physical quantities—such as energy, angular momentum, and action—occur in finite units rather than varying continuously. This discreteness is governed by **Planck's constant**, which sets the scale at which quantum effects become unavoidable.

At scales approaching the **Planck length and Planck time**, even the standard quantum description may become incomplete. These scales lie far beyond current experimental reach and suggest a regime where spacetime itself may no longer be smooth or continuous. Whether this represents a deeper quantum layer or a fundamentally different description of reality remains an open question.

Within this book, the quantum is treated not merely as a microphysical theory, but as a **shift in how reality is understood**. It replaces certainty with structure, objects with relations, and determinism with constrained possibility.

The quantum framework reveals that reality is not built from fully formed things, but from **potentialities governed by precise rules**—rules that allow coherence, pattern, and emergence to arise without classical predictability.

In this sense, the quantum does not represent the final layer of reality, but a **threshold**: a domain where classical description dissolves, and deeper questions about spacetime, information, and emergence begin.

Appendix C.10—Information

Information refers to structured distinction—differences that make a difference within a system.

In physical sciences, information is not merely data or symbols, but a measure of **organisation, constraint, and correlation**. It describes how states relate, how possibilities are limited, and how structure is preserved or transformed across space and time.

Within this book, information is treated as a **foundational substrate of reality**, alongside space, time, and fields. Matter, energy, and physical laws may be understood as expressions of how information is encoded, conserved, and propagated in the Universe.

Information is **not equivalent to intelligence or consciousness**. Rather, it provides the conditions under which intelligence may arise—by enabling pattern formation, memory, and relational coherence.

Importantly, information does not require awareness to exist. It precedes interpretation, meaning, and understanding. Only when information is organised, integrated, and acted upon does it participate in higher-order processes such as cognition, learning, or wisdom.

In this sense, information occupies a **middle layer**:

below experience and awareness,

above randomness and noise,

and essential to emergence across physical, biological, and cognitive systems.

Appendix C.11—Emergence (including emergent time)

Emergence refers to the appearance of coherent patterns, structures, or behaviours that arise from interactions among simpler components, yet are not predictable by examining those components in isolation. What emerges is not added from outside, but becomes visible only at the level of collective organisation.

In physical systems, emergence explains how temperature arises from molecular motion, how traffic jams form without a central controller, or how biological order develops from chemical processes. The governing laws at the lower level remain intact, but new descriptions become necessary at higher levels.

Within this book, **emergence is treated as a structural principle rather than a metaphysical claim**. It marks a transition in description—from parts to relations, from local rules to global patterns.

Emergent time extends this idea to temporality itself. Rather than assuming time as a fundamental background against which events occur, some contemporary approaches suggest that temporal ordering may arise from correlations, information flow, or causal structure within physical systems. In this view, time is not eliminated, but understood as a *derived feature* of deeper relational dynamics.

This does not deny lived experience or measured duration. Instead, it reframes time as something that becomes meaningful through interaction, change, and organisation—much as space gains structure through fields and matter.

Emergence, including emergent time, thus supports a central theme of this inquiry: that reality may be **less a collection of primitive absolutes and more a layered coherence**, where meaning and structure arise together.

Appendix D—Conceptual Visualisations (Textual descriptions)

D.1 Fields of Modern Physics

D.2 Spacetime Curvature

D.3 Black Hole Light-Cone Geometry

D.4 Quantum Measurement Flow

D.5 Quantum Interference and the Emergence of Outcomes

D.6 Multiverse Bubble Landscape

D.7 Chirality and Mirror Asymmetry

D.8 Higgs Field vs Higgs Boson

D.9 The Layers of Reality: Domains and Dominant Descriptions

Appendix D.1—Fields of Modern Physics

This section presents a layered view of the major fields of modern physics, illustrating how different theoretical frameworks operate at different scales of reality.

At the macroscopic scale, classical mechanics and classical field theories describe everyday motion, forces, and electromagnetic phenomena. These frameworks work extremely well where quantum effects are negligible.

At smaller scales, quantum mechanics and quantum field theory describe the behaviour of matter and interactions at the level of particles and fields. Here, probability amplitudes, superposition, and quantisation replace classical determinism.

At the cosmic scale, general relativity governs the structure of spacetime itself, explaining gravity as curvature rather than force. It accurately describes black holes, gravitational waves, and the large-scale evolution of the universe.

An important limitation is highlighted by this layered view: these frameworks coexist but are not yet unified. Quantum field theory excels at small scales, while general relativity governs large-scale structure. Their overlap—where both quantum effects and strong gravity are significant—remains an open frontier of physics.

Within the context of this book, this framing serves two purposes:

> to show that no single theory explains all of reality, and
>
> to reinforce the idea that reality is best understood as layered, relational, and context-dependent.

This is not a claim of progress from “lower” to “higher” truth, but a reminder that different descriptions are valid within different domains—each revealing structure while leaving deeper questions open.

Appendix D.2—Spacetime Curvature

This conceptual visualisation describes spacetime not as a passive backdrop in which events occur, but as a dynamic geometric structure whose curvature is shaped by mass and energy.

In general relativity, gravity is not treated as a force acting across space. Instead, massive objects alter the geometry of spacetime itself. Objects and light then follow paths—called geodesics—determined by this curved geometry. What appears as gravitational attraction is, in this view, motion along the natural contours of spacetime.

A common way to introduce this idea is by analogy, imagining spacetime as a stretched surface that bends underweight. While such analogies can be helpful, they must be treated with care. Spacetime is not a physical sheet embedded in a higher-dimensional space. Curvature is an intrinsic property of spacetime geometry, defined mathematically rather than visually.

This description emphasises a key conceptual shift: motion under gravity does not require a pulling force. Instead, trajectories emerge from the structure of spacetime itself. Light bending near massive objects, time running at different rates in different gravitational fields, and the dynamics of orbiting bodies all arise from this geometric framework.

Spacetime curvature thus represents a unification of space, time, and gravitation into a single relational structure. Matter and energy determine how spacetime is shaped, and spacetime geometry, in turn, determines how matter and energy move.

This framing prepares the ground for later ideas in the inquiry, where structure, relation, and constraint play a central role in explaining how physical behaviour emerges without invoking additional forces or hidden mechanisms.

Appendix D.3—Black Hole Light-Cone Geometry

This conceptual visualisation describes how causality and the flow of time are shaped by extreme spacetime curvature in the vicinity of a black hole.

Light cones represent the set of all possible directions in spacetime that light—and therefore causal influence—can take from a given event. They define what counts as the future, the past, and the elsewhere (regions that cannot be causally connected).

Far from a black hole, light cones are oriented symmetrically. Under such conditions, motion and communication are possible both toward and away from massive objects. As spacetime curvature increases near a black hole, the orientation of light cones progressively tilts inward.

At the event horizon, this tilting becomes so extreme that all future-directed paths—including those followed by light—point inward. Beyond this boundary, no signal can escape to the outside Universe. The event horizon therefore marks a causal boundary, not a physical surface or barrier.

A crucial conceptual point follows from this structure: falling into a black hole does not involve crossing a force wall or encountering a singular obstruction. Instead, it involves entering a region where spacetime geometry itself redefines what "forward in time" means. Inside the horizon, moving outward is no longer a possible future direction, just as moving backward in time is impossible outside it.

Within the context of this book, this visualisation illustrates how causality, time, and geometry are inseparably linked. Extreme conditions do not violate physical law; rather, they reveal its full implications when structure and constraint dominate behaviour.

Appendix D.4—Quantum Measurement Flow

This section presents a conceptual account of how quantum systems transition from multiple possible states to a single observed outcome.

At the initial stage, a quantum system is described by a wavefunction encoding a range of possible states. These possibilities coexist as probability amplitudes rather than as definite properties. No single outcome is privileged at this stage.

When the system interacts with a measuring apparatus or with its surrounding environment, this interaction constrains the system. Information about the quantum state becomes distributed across many degrees of freedom. As a result, interference between alternative possibilities becomes effectively inaccessible.

This transition can be understood as progressing:

from superposition, through interaction and entanglement, to a definite recorded outcome.

Importantly, this description does not assign a special causal role to consciousness. Measurement is treated as a physical interaction and an update of information, consistent with standard quantum interpretations discussed elsewhere in the book.

Within the context of this inquiry, measurement is emphasised not as an instantaneous or mysterious event, but as a process—one in which possibility gives way to actuality through interaction, constraint, and information flow.

This conceptual framing clarifies how classical definiteness can emerge from quantum structure without invoking hidden variables, additional forces, or metaphysical assumptions.

Appendix D.5—Quantum Interference And The Emergence Of Outcomes

This section presents a conceptual visualisation of how quantum interference shapes physical outcomes before any measurement occurs.

At the quantum level, a system does not evolve along a single definite path. Instead, it evolves through a space of allowed possibilities, each associated with a probability amplitude rather than a classical probability. These amplitudes coexist simultaneously and interact with one another, even when the system is isolated and unobserved.

At the level of possibility space, multiple configurations, paths, or states are permitted. None of these represents an outcome; they represent structured potential. At this stage, nothing has yet occurred in the classical sense.

As the system evolves, interference takes place. Probability amplitudes combine—some reinforcing one another, others cancelling out. This process occurs prior to observation and independent of measurement. It is here that structure emerges. Certain possibilities become strongly favoured, while others are suppressed or eliminated—not by chance, but by the relational geometry of the amplitudes themselves.

Only after this interference has shaped the landscape of possibilities does measurement occur. At that point, probabilities—given by the squared magnitudes of the resulting amplitudes—govern which outcome is realised. What appears as a single, definite event is thus the final expression of a deeper, structured process.

The conceptual significance of this process is central to quantum theory: reality does not arise from isolated possibilities, but from how possibilities interact.

Quantum interference explains why outcomes are neither purely random nor strictly predetermined. They are constrained, patterned, and selective even before observation. Within the broader inquiry of this book, quantum interference stands as a clear example of emergence—order arising not from certainty, but from structured potential.

Appendix D.6—Multiverse Bubble Landscape

This section presents a conceptual visualisation of the multiverse as a landscape of possible Universes, rather than as a collection of physically observable worlds.

In several modern cosmological frameworks—particularly those emerging from inflationary theory and high-energy physics—our observable universe may be understood as one region within a much larger cosmic structure. In this view, different regions of spacetime can settle into distinct configurations of physical laws, constants, or vacuum states, giving rise to multiple possible "bubble universes."

Bubble Formation

Each bubble represents a region in which spacetime has stabilised into a particular configuration of physical parameters. Such regions may arise from an underlying inflationary background or deeper vacuum structure. Once formed, a bubble evolves largely independently, with its own internal history and causal order.

These bubbles are not embedded within ordinary space. They are not separated by distance in the conventional sense and cannot be navigated between using familiar notions of motion or travel. Their separation is structural and causal, not spatial.

Observable Limits

From within any single bubble—including our own—other bubbles are, in principle, unobservable. There is no exchange of signals, no shared temporal reference, and no causal interaction once such regions separate. As a result, multiverse models remain theoretical extensions, inferred indirectly from mathematical consistency and explanatory reach rather than direct observation.

Conceptual Role

Within the context of this book, the multiverse is not presented as a claim about what exists, but as a boundary concept. It illustrates how modern physics increasingly describes reality not as a single inevitable outcome, but as a structured space of possibilities constrained by deeper principles.

Whether or not multiple Universes exist, this framework highlights a significant shift in perspective: physical law may allow a range of coherent realisations rather than dictate a unique form.

Relation to Emergence

Seen in this way, the multiverse resonates with themes explored elsewhere in the inquiry—emergence, constraint, and selection. Our Universe may be understood not as an arbitrary accident, but as one coherent realisation among many potential ones, shaped by underlying structure rather than pure chance.

Appendix D.7—Chirality And Mirror Asymmetry

This section conceptually illustrates chirality—the property by which an object and its mirror image cannot be superimposed—and its profound role in the structure of reality.

At first glance, many physical systems appear symmetric. However, at deeper levels, nature consistently exhibits *broken symmetry*, where left and right are no longer interchangeable. Chirality captures this distinction.

Mirror Asymmetry

A chiral object differs fundamentally from its mirror image. The two forms may appear identical in shape, yet cannot be aligned through rotation or translation. Human hands provide the most familiar example, but chirality extends far beyond anatomy.

In physics and chemistry, chirality is found in:

> the handedness of fundamental particles under the weak interaction,
>
> the consistent left-handedness of amino acids in biological systems,
>
> the right-handed structure of DNA.

These asymmetries are not random irregularities; they are *systematic and universal.*

Chirality as a Structural Principle

Chirality demonstrates how **asymmetry enables structure**. A perfectly symmetric universe would lack direction, distinction, and differentiation. Chirality introduces orientation—a "this way, not that way"—allowing processes to unfold with coherence and continuity.

In this sense, chirality acts as a bridge between abstract symmetry and concrete form.

Conceptual Significance

Within the broader inquiry of this book, chirality exemplifies a recurring insight:

Symmetry defines what is possible; asymmetry determines what becomes real.

This conceptual visualisation highlights how small, foundational asymmetries can cascade upward, shaping complexity in physics, chemistry, biology, and ultimately life itself.

Chirality reminds us that the Universe is not merely governed by balance, but by *selective imbalance*—a subtle bias that makes identity, evolution, and direction possible.

Appendix D.8—Higgs Field Vs Higgs Boson

This section clarifies a distinction that is often misunderstood but conceptually crucial: the difference between a *field* and its *excitation*.

The Higgs Field

The Higgs field is a continuous, invisible field that permeates all of space. It is always present, even in regions devoid of particles. In this sense, it functions not as an object but as a background condition of reality.

Particles acquire mass not because the Higgs field acts upon them like a force, but because their interaction with this field constrains how they move through spacetime. Mass, from this perspective, is not a substance but a relational property arising from field interaction.

The Higgs Boson

The Higgs boson is not the field itself. It is a localised excitation—a temporary ripple or disturbance in the Higgs field—detectable only under extreme energy conditions, such as those created in particle accelerators.

Its fleeting existence is precisely what makes it observable. The boson serves as empirical evidence of the field's reality, not as its source.

Conceptual Significance

This distinction reflects a broader pattern found throughout modern physics:

Field—enduring, structural, omnipresent

Particle—transient, localised, observable

Just as electromagnetic waves reveal the electromagnetic field, the Higgs boson reveals the Higgs field. The field is fundamental; the particle is its expression.

Why This Matters for the Inquiry

This clarification reinforces a central theme of the book: reality is structured more by invisible frameworks than by visible objects.

What we detect are moments of manifestation. What governs behaviour are persistent, unseen fields. The Higgs field exemplifies how form, resistance, and identity can arise from an underlying relational medium rather than from material substance.

Appendix D.9—The Layers Of Reality: Domains And Dominant Descriptions

Reality is not best understood as a single homogeneous level governed by one kind of law. Instead, it reveals itself as a layered structure, where different descriptions become meaningful at different scales and contexts.

These layers are not stacked like physical floors, nor do they represent separate worlds. They are **domains of effective description**, each governed by its own dominant principles, constraints, and explanatory tools.

Foundational Physical Domain

At the most fundamental level lie quantum fields and spacetime structure. Here, reality is described not in terms of objects, but in terms of amplitudes, fields, symmetries, and constraints. Causality is no longer strictly classical, and classical intuitions such as locality and determinism lose their primacy.

This domain sets the **boundary conditions** for all higher layers, even though its concepts are rarely visible in everyday experience.

Emergent Physical Domain

From these foundations emerge stable particles, forces, and classical spacetime. At this level, objects, trajectories, and causal chains become meaningful approximations. Classical physics works extraordinarily well here—not because it is fundamental, but because it is **structurally sufficient** at this scale.

Emergence does not negate the underlying quantum reality; it constrains and stabilises it.

Biological Domain

Life introduces a new layer where organisation, metabolism, and self-maintenance dominate. The relevant questions shift from “What are the laws?” to “How is stability preserved far from equilibrium?”

Biological systems obey physical laws but cannot be explained by them alone. Information, regulation, and adaptation become indispensable concepts.

Cognitive and Conscious Domain

With nervous systems and minds, reality acquires perspective. Meaning, intention, memory, and interpretation appear—not as violations of physics, but as emergent relational properties of complex systems.

At this layer, observation is no longer passive. The observer participates in shaping experience, interpretation, and understanding.

Social and Symbolic Domain

Human language, culture, mathematics, science, and ethics constitute another layer of reality. Here, symbols, norms, institutions, and shared meanings exert causal influence, even though they are not physical entities.

This domain demonstrates how **non-material structures** can guide behaviour, decision-making, and collective evolution.

Why "Layers" Matter

Each layer:

depends on lower layers,

is constrained by them,

yet introduces **novel explanatory concepts** irreducible to those below.

No single layer has explanatory monopoly. Attempting to reduce higher layers entirely to lower ones leads to category errors; treating higher layers as independent of lower ones leads to mystification.

Insight for the Inquiry

The layered view of reality supports a central conclusion of this book:

Reality is not fragmented, but **plural in description**.

Different domains require different lenses, not because truth is relative, but because structure expresses itself differently across scales. Understanding arises not from collapsing these layers into one, but from recognising how they cohere.

Appendix E—Cosmic Probability And Seti Framework

The frameworks discussed here are not analyses of SETI programs themselves, but of the conceptual and probabilistic foundations required for any search for extraterrestrial intelligence.

E.1 The Drake Equation

E.2 The Fermi Paradox

E.3 The Great Filter Hypothesis

E.4 Cosmic Probability Landscapes

E.5 Extremophiles and Planetary Analogs

E.6 SETI Detection and Interpretation Framework

Appendix E.1—The Drake Equation

The Drake Equation is not a predictive formula, but a **conceptual framework** introduced by astronomer **Frank Drake** to organise thinking about the possibility of extraterrestrial intelligent life.

Rather than answering *whether* such life exists, it asks a more careful question:

What factors must align for a communicating civilisation to be detectable at all?

The Structure of the Equation

In simplified form, the Drake Equation multiplies several factors:

- the rate of star formation
- the fraction of stars with planets
- the number of habitable planets per system
- the likelihood that life arises
- the likelihood that intelligence emerges
- the likelihood that intelligence becomes technological
- the duration for which such civilisations remain detectable

Each term represents a **filter**, not a certainty.

Crucially, most of these quantities are **not independent** and **not directly measurable**. They span different domains: astrophysics, chemistry, biology, evolution, technology, and sociology.

Why the Drake Equation Matters

The lasting value of the Drake Equation lies in what it *does not* do.

It does not claim:

- that life is common
- that intelligence is inevitable
- that civilisations endure
- or that detection is likely

Instead, it **forces structure onto ignorance**.

By laying out the dependencies explicitly, the equation clarifies where uncertainty truly lies—and how small changes in assumptions can lead to radically different conclusions.

In this sense, the Drake Equation is an **exercise in disciplined humility**.

A Relational, Not Deterministic, Tool

Within the context of this book, the Drake Equation is best understood as a **relational model**:

- It treats intelligence as an **emergent outcome**, not a guaranteed endpoint.
- It recognises that cosmic possibility is shaped by **constraints, bottlenecks, and lifetimes.**
- It highlights that detection depends as much on **overlap in time** as on existence itself.

Even if intelligent life arises frequently, the chances of **simultaneous detectability** may remain vanishingly small.

Position Within This Inquiry

The Drake Equation aligns with the broader themes of this inquiry:

Possibility precedes actuality

Structure precedes certainty

Emergence is shaped by filters, not destiny

It serves as a bridge between **cosmic scale** and **methodological restraint**, preparing the ground for deeper questions explored in the following sections—particularly the **Fermi Paradox** and the **Great Filter Hypothesis**.

The equation does not tell us where others are.

It tells us **why the question itself is so difficult.**

Appendix E.2—The Fermi Paradox

The Fermi Paradox arises from a simple tension between expectation and observation.

On one hand, the vastness of the Universe—with its billions of galaxies, trillions of stars, and immense timescales—suggests that intelligent life **should** be common. On the other hand, despite decades of searching, we observe **no clear evidence** of extraterrestrial civilisations.

This contrast is often summarised in a single question attributed to physicist **Enrico Fermi**:

"Where is everybody?"

Why It Is a Paradox

The paradox does not stem from a calculation error, but from a **conceptual mismatch**.

If even a small fraction of intelligent civilisations develop interstellar capability, then—given enough time—their presence should be detectable through artifacts, signals, or large-scale cosmic engineering.

Yet the observable universe appears **quiet**.

No probes, no beacons, no unmistakable signatures.

The paradox lies not in absence alone, but in the **discrepancy between plausible abundance and apparent silence**.

Common Classes of Explanations

Over time, explanations for the Fermi Paradox have fallen into several broad categories:

Rarity: Intelligent life may be extraordinarily rare.

Fragility: Civilisations may self-destruct or stagnate quickly.

Isolation: Civilisations may exist but remain technologically or communicatively isolated.

Invisibility: Advanced intelligence may not emit detectable signals or may use modes beyond our recognition.

Timing: Civilisations may arise at different epochs, rarely overlapping in detectable windows.

Each explanation shifts the tension but does not dissolve it entirely.

A Deeper Reframing

Within this inquiry, the Fermi Paradox is not treated as a failure of expectation, but as a **lesson in projection**.

It reveals a common assumption: that intelligence naturally seeks expansion, visibility, or permanence.

Yet intelligence, as explored throughout this book, is an **emergent, context-dependent phenomenon**, shaped by constraints, environments, and internal priorities—not a universal trajectory toward cosmic presence.

Silence, therefore, may not imply absence.

It may imply **non-convergence**.

Relational Silence

The paradox also highlights the role of **relational conditions**:

Overlap in time

Overlap in communication modality

Overlap in interpretive frameworks

Without alignment across these dimensions, coexistence does not guarantee contact.

In this sense, the Universe may not be empty—only **uncoordinated**.

Position Within This Inquiry

The Fermi Paradox serves as a conceptual hinge in this appendix.

The Drake Equation structures *possibility*

The Fermi Paradox confronts *expectation*

The next step examines *filters*

Rather than demanding resolution, the paradox invites restraint.

It cautions against extrapolating from **human narratives** to **cosmic outcomes**, and prepares the ground for the more sobering question addressed next:

If intelligence is possible, what prevents it from becoming enduring and visible?

That question leads directly to the **Great Filter Hypothesis.**

Appendix E.3—The Great Filter Hypothesis

The **Great Filter Hypothesis** is an attempt to explain the silence highlighted by the Fermi Paradox by introducing a sobering possibility:

Some step between simple matter and detectable cosmic intelligence is extraordinarily difficult to pass.

This step—or sequence of steps—is referred to as the *Great Filter*.

What the Filter Represents

The Great Filter is not a single event, nor is it necessarily catastrophic. It may consist of one or more **low-probability transitions** in the long chain from cosmic beginnings to technological civilisation.

Commonly proposed stages include:

- The emergence of life from non-life
- The transition from simple to complex organisms
- The development of intelligence
- The rise of technological civilisation
- Long-term survival without self-destruction
- Expansion or detectability at cosmic scales

The hypothesis does not assert which stage is the filter—only that **at least one stage must be exceptionally rare**, given the observed absence of visible civilisations.

Past or Future?

The unsettling power of the Great Filter lies in a single question:

Is the filter behind us—or ahead of us?

If the hardest steps lie **in the past**, then humanity may be unusually fortunate.

If the hardest steps lie **in the future**, then extinction, stagnation, or invisibility may be the typical fate of technological intelligence.

The hypothesis offers no comfort—only orientation.

Not a Prediction, but a Constraint

Within this inquiry, the Great Filter is treated not as a prophecy, but as a **constraint-based reasoning tool**.

It reframes the problem from *"Why don't we see others?"* to:

"Which assumptions about progress, survival, and visibility are unjustified?"

The filter may not eliminate intelligence outright. It may instead limit **duration, coordination**, or **detectability**.

Beyond Catastrophe

Importantly, the Great Filter does not require dramatic annihilation.

Subtle filters may include:

Loss of curiosity or exploratory drive

Technological saturation without expansion

Preference for inward or simulated realities

Ethical restraint against interference

Communication modes incompatible with detection

In such cases, civilisations may persist—yet leave no traceable cosmic footprint.

Relevance to This Book

Throughout this work, intelligence is treated as **emergent, constrained, and relational**, not as a guaranteed ladder toward dominance.

The Great Filter aligns with this view by emphasizing:

Non-linearity of development

Fragility of coordination

Dependence on environmental and internal balance

It cautions against assuming that intelligence naturally scales.

A Mirror, Not a Threat

Finally, the Great Filter acts as a mirror.

It does not tell us what *will* happen—only what *must be accounted for.*

Whether humanity stands before or beyond the filter remains unknown. But the hypothesis reinforces a central theme of this inquiry:

The Universe permits intelligence—it does not promise its endurance.

This sets the stage for the next question:

If intelligence is constrained rather than absent, **how should probability itself be understood across cosmic scales?**

That question leads naturally to **cosmic probability landscapes.**

Appendix E.4—Cosmic Probability Landscapes

When considering intelligence at cosmic scales, probability cannot be treated as a single number or a simple likelihood. Instead, it must be understood as a **landscape**—shaped by multiple variables, constraints, and dependencies that interact across vast scales of space and time.

A *cosmic probability landscape* describes how outcomes are distributed not uniformly, but unevenly, across the Universe.

Probability as Structure, Not Chance

In everyday contexts, probability is often associated with randomness or uncertainty. In cosmic contexts, however, probability reflects **structure**:

physical laws

environmental constraints

evolutionary pathways

temporal windows

survival durations

These factors do not act independently. They combine to form regions of **high possibility** and **near impossibility**, much like peaks and valleys in a landscape.

Intelligence, within this view, is not "unlikely by chance"—it is **conditional by structure**.

Temporal Narrowness

One of the most decisive features of the cosmic probability landscape is **time**.

Even if intelligent civilisations arise frequently, their periods of technological detectability may be:

brief

non-overlapping

unstable

Probability, therefore, is not merely spatial ("How many?"), but **spatiotemporal** ("When, and for how long?").

Two civilisations separated by millions of years may both exist—yet never coincide within a shared window of detectability.

Multiplicity Without Convergence

A landscape can contain many peaks without any two peaks touching.

Likewise, the Universe may host many forms of intelligence that:

arise under different conditions

follow different technological paths

prioritise different modes of existence

communicate in incompatible ways

From this perspective, the absence of contact does not imply emptiness. It implies **non-convergence** within the probability landscape.

Human Position Within the Landscape

Humanity occupies a very narrow ridge within this vast terrain:

a specific planetary environment

a brief technological phase

a fragile coordination of biology, culture, and tools

Our current detectability—through radio emissions, atmospheric signatures, or artifacts—may be neither typical nor durable.

What appears central from our vantage point may be **statistically marginal** at cosmic scale.

Alignment With This Inquiry

Throughout this book, reality is approached as structured by **constraints rather than guarantees**.

Cosmic probability landscapes reflect this same logic:

outcomes emerge where conditions align

rarity does not imply impossibility

abundance does not ensure visibility

Intelligence, like many complex phenomena, may occupy **small but real regions** of a vast possibility space.

The Universe may be rich in potential—yet sparse in overlap.

Transition Forward

Seen this way, cosmic silence is not a mystery demanding resolution, but a signal demanding restraint.

Probability landscapes do not answer whether others exist. They clarify **why expectation must be carefully bounded**.

This prepares the ground for the next consideration:

How do extreme environments—and the adaptability of life itself—reshape what we consider possible?

That question leads naturally to **extremophiles and planetary analogs**.

Appendix E.5—Extremophiles And Planetary Analogs

For much of human history, life was assumed to require a narrow set of conditions: moderate temperatures, liquid water, stable atmospheres, and Earth-like chemistry. Modern biology has steadily dismantled this assumption.

The discovery of **extremophiles**—organisms that thrive in environments once thought inhospitable to life—has profoundly expanded our understanding of biological possibility.

Life at the Edge of Conditions

Extremophiles are found in environments characterised by extremes of:

- temperature (near boiling or freezing)
- pressure (deep oceans)
- acidity or alkalinity
- salinity
- radiation
- chemical toxicity

Rather than merely surviving these conditions, such organisms often depend on them.

This reveals an important insight:

Life adapts to constraints rather than requiring ideal conditions.

Redefining Habitability

The existence of extremophiles forces a revision of what "habitable" means.

Habitability is no longer defined by resemblance to Earth's surface conditions, but by the presence of:

stable energy gradients

chemical disequilibrium

sustained environmental structure

Liquid water remains significant, but it is not the sole criterion. In some cases, life exploits chemical reactions entirely independent of sunlight.

Habitability, therefore, becomes **contextual rather than universal**.

Planetary Analogs

Extremophile research directly informs the search for life beyond Earth through the concept of **planetary analogs**.

Environments on Earth are studied as partial stand-ins for extraterrestrial conditions, such as:

subsurface oceans beneath icy crusts

high-radiation surfaces

methane-rich atmospheres

chemically reactive subsurfaces

These analogs do not imply similarity in organisms, but similarity in **functional constraints**.

The question shifts from *"Is this planet Earth-like?"* to:

"Does this environment sustain persistent structure against entropy?"

Implications for Cosmic Probability

Extremophiles expand the **possibility space** for life—but they do not guarantee intelligence.

Life may be robust across environments, while intelligence remains rare, fragile, or transient.

In this sense, extremophiles widen the base of the probability landscape without necessarily raising its highest peaks.

Alignment With This Inquiry

Throughout this book, complexity arises not from ideal conditions, but from **structured constraint.**

Extremophiles exemplify this principle biologically:

They show that life is not narrowly tuned

They demonstrate adaptability without inevitability

They reinforce that emergence is opportunistic, not designed

Life appears wherever conditions allow **persistent organisation**, not wherever conditions resemble our own.

Looking Ahead

If life can arise and persist under such diverse *and adverse* conditions, the absence of visible intelligence elsewhere cannot be explained solely by environmental hostility.

The challenge shifts again:

If life is adaptable, **what limits the emergence, coordination, or longevity of intelligence itself?**

That question prepares the ground for the final section of this appendix, which examines how detection, interpretation, and expectation shape our cosmic search.

Appendix E.6—Seti Detection And Interpretation Framework

The Search for Extraterrestrial Intelligence (SETI) is often framed as a technological endeavor—scanning the sky for signals, anomalies, or artifacts. In this inquiry, however, SETI is treated **not as a program or prediction**, but as a **methodological framework** for reasoning about detectability, interpretation, and cosmic silence.

SETI highlights a crucial distinction: the difference between **existence, detectability**, and **interpretability.**

A civilisation may exist without being detectable. A detectable signal may exist without being interpretable. And interpretability itself depends on shared structures—physics, information encoding, and constraints—rather than intent alone.

From this perspective, SETI becomes a test case for a broader principle explored throughout this book: **reality is structured by constraints, not guaranteed by possibility.**

Detection is not guaranteed by existence

Even technologically advanced civilisations may leave no traceable cosmic footprint. Signals may be transient, deliberately quiet, indistinguishable from noise, or encoded in ways inaccessible to our current conceptual frameworks.

Thus, absence of evidence does not imply absence of life or intelligence—only the absence of detectable interaction under present constraints.

Interpretation is constrained by shared structure

For a signal to be recognised as artificial, it must conform to patterns distinguishable from natural processes. This requires shared assumptions about mathematics, physics, symmetry, and information—assumptions that may not hold universally.

SETI therefore exposes the limits of anthropocentric reasoning: detection depends not only on *what is sent*, but on *what can be recognised*.

Methodological significance

Within this appendix, SETI functions as a **discipline of restraint**. It cautions against:

Overinterpreting silence,

Confusing technological capability with inevitability,

Treating probability as destiny.

Instead, it reinforces a core theme of this inquiry: cosmic outcomes emerge from layered constraints—physical, informational, and contextual—rather than from raw possibility alone.

In this sense, SETI is less about finding others, and more about understanding the conditions under which **any reality becomes observable at all.**

Appendix F—Methodological Notes

The following notes clarify the methodological stance underlying this inquiry. They are not intended to be read sequentially, but as complementary lenses applied where appropriate.

F.1 Systems Thinking Approach

F.2 Boundary Models (Classical ↔ Quantum)

F.3 Using Metaphors in Scientific Thinking

F.4 Emergence and Complexity Reasoning

F.5 Avoiding Category Errors

F.6 Integrating Vedanta and Physics (methodological discipline)

F.7 Critical Filters for Evaluating Scientific Claims

F.8 Zero, Nothingness, and Emptiness: Conceptual Clarifications

F.9 Consciousness as a Distributed Phenomenon

F.10 Structure, Relation, and Constraint: A Relational Methodological Stance

Appendix F.1—Systems Thinking Approach

This inquiry adopts a **systems thinking approach** to understand reality not as a collection of isolated entities, but as an interconnected whole governed by structure, relation, and constraint.

From Objects to Systems

Classical intuition often begins with objects and asks how they interact. Systems thinking reverses this orientation:

It begins with **relationships**, not things

It emphasises **patterns of interaction**, not isolated properties

It treats behaviour as an outcome of **system structure**, not merely local causes

In this view, entities derive their meaning from the systems they participate in, rather than possessing complete, independent definitions.

Emergence over Reduction

Systems thinking recognises that:

Complex behaviour can arise from simple rules

Global properties need not be present in individual components

Reduction explains components, but not always the behaviour of the whole

This perspective is essential for understanding phenomena such as emergence, self-organisation, phase transitions, and collective order—all of which appear repeatedly across physics, biology, cognition, and cosmology.

Constraints as Generative, Not Limiting

Within a system, constraints are not merely restrictions; they are **enabling conditions**. They shape what is possible, stable, and observable.

Examples include:

Physical constants constraining allowable structures

Conservation laws shaping dynamics

Boundary conditions defining system behaviour

In this inquiry, constraints are treated as **structural features of reality**, not arbitrary limitations imposed from outside.

Multi-Scale Coherence

Systems thinking allows consistent reasoning across scales:

Quantum → classical

Local → global

Micro → macro

Rather than forcing explanations at a single level, this approach accepts that different descriptions may be valid at different scales—provided they remain coherent with one another.

Why This Approach Is Chosen

This methodological stance is adopted because it:

Avoids category errors caused by isolating phenomena prematurely

Preserves coherence across domains without reductionism

Aligns naturally with modern physics, complexity science, and relational interpretations

Supports inquiry without premature metaphysical closure

Throughout this book, systems thinking serves not as a doctrine, but as a **discipline of attention**—guiding how questions are framed, how explanations are evaluated, and how limits are respected.

Appendix F.2—Boundary Models (Classical ↔ Quantum)

Much of modern scientific difficulty arises not from errors within theories, but from confusion at the **boundaries between domains.** Classical physics and quantum physics are each internally coherent, yet they operate under fundamentally different assumptions about determinism, measurement, and description. Problems emerge when concepts native to one domain are uncritically extended into the other.

Boundary models are conceptual tools used to **clarify where one descriptive regime gives way to another**, without demanding premature unification. They acknowledge that classical behaviour emerges from quantum systems under certain conditions—such as scale, decoherence, and environmental interaction—while also recognising that quantum descriptions cannot simply be reduced to classical intuition.

In this inquiry, boundary models serve a **disciplinary role** rather than a unifying one. They prevent category errors by reminding the reader that explanations are context-dependent: what counts as an object, a cause, or a measurement differs across domains. Rather than asking which description is "more real," boundary thinking asks **which description is appropriate to the scale and question at hand.**

This approach resists both reductionism and mystification. It neither collapses classical reality into quantum abstraction nor treats quantum phenomena as anomalies within a classical worldview. Instead, it preserves the integrity of each domain while allowing for coherent transition between them.

These boundaries are not merely theoretical—they are encountered in ongoing experimental physics. Contemporary experiments continue to reveal small but persistent deviations from established models, often described as "hints of new physics." Within the framework of this inquiry, such signals are not treated as disruptions of reality itself, but as indicators of the limits of current descriptive regimes. What is revealed in such moments is not a breakdown of structure, but a boundary of articulation. The underlying constraints remain coherent; it is our models that require extension.

By respecting boundaries without hardening them into absolutes, boundary models enable inquiry to proceed with clarity, humility, and conceptual precision.

Appendix F.3—Using Metaphors In Scientific Thinking

Metaphors play an unavoidable role in human understanding, especially when inquiry moves beyond direct sensory experience. Concepts such as *fields*, *waves*, *particles*, *fabric of spacetime*, or *information flow* are not literal descriptions of reality but **conceptual bridges** that allow abstract structures to be grasped intuitively.

In scientific thinking, metaphors are most valuable when they are treated as **tools of insight rather than claims of identity**. A metaphor can illuminate relationships, constraints, or dynamics without asserting that reality itself is composed of the metaphorical elements invoked. Confusion arises when metaphors are unconsciously reified—when explanatory aids are mistaken for ontological facts.

This inquiry uses metaphors deliberately and sparingly. Their purpose is not to simplify complexity into comfort, but to **render structure visible without distorting it**. When used well, metaphors guide intuition toward coherence; when used carelessly, they introduce hidden assumptions and category errors.

A disciplined approach to metaphor requires continual awareness of its limits. Every metaphor highlights certain aspects of a phenomenon while obscuring others. Recognising where a metaphor ceases to apply is as important as appreciating where it succeeds. In this sense, metaphors function best when they remain **transparent**, allowing the underlying structure to be seen through them rather than replaced by them.

By maintaining this discipline, metaphors can support inquiry without constraining it—clarifying understanding while preserving the openness essential to deep scientific and philosophical exploration.

Appendix F.4—Emergence And Complexity Reasoning

Emergence refers to the appearance of coherent patterns, behaviours, or properties at a collective level that are **not explicitly present in the individual components.** Complexity reasoning recognises that such patterns arise not through central control or linear causation, but through interactions governed by local rules and constraints.

In this inquiry, emergence is treated neither as a mystery nor as a shortcut explanation. It is understood as a **structural outcome**: when many elements interact within defined constraints, higher-level order can arise that requires its own descriptive language. This does not deny the relevance of lower-level processes, but acknowledges that explanation must match the level at which coherence appears.

Complex systems often display sensitivity to initial conditions, nonlinear responses, and adaptive behaviour. These features make prediction difficult, but they do not imply randomness. Rather, they reflect the fact that **global behaviour is shaped by relational structure**, not by isolated components acting independently.

Emergence reasoning helps avoid two common errors. The first is reductionism, which assumes that explaining parts automatically explains wholes. The second is mystification, which treats emergent phenomena as inexplicable or irreducible in principle. A disciplined approach recognises emergence as **context-dependent order**, requiring explanation at the appropriate scale.

By applying emergence and complexity reasoning, this inquiry remains attentive to how structure, relation, and constraint give rise to form across domains—from physical systems to life, cognition, and collective intelligence—without collapsing distinct levels of description into one another.

Appendix F.5—Avoiding Category Errors

A category error occurs when a concept appropriate to one domain of description is mistakenly applied to another. Such errors do not arise from incorrect facts, but from **misplaced questions**—asking something of reality that the chosen conceptual framework cannot meaningfully answer.

In scientific and philosophical inquiry, category errors often appear when language, intuition, or metaphor is carried across boundaries without sufficient scrutiny. Examples include treating probability as ignorance rather than structure, assuming classical causality in quantum contexts, or attributing intentionality to impersonal physical processes. These errors persist because the concepts involved feel familiar, even when they are inappropriate.

This inquiry places particular emphasis on avoiding category errors because many foundational debates—about time, consciousness, reality, or causation—are sustained not by empirical disagreement, but by **conceptual misalignment**. When categories are confused, debates become circular, and progress appears blocked.

Avoiding category errors requires continual attention to the **level and domain of explanation** being employed. Questions must be matched to the descriptive framework capable of addressing them. This does not imply rigid separation between domains, but disciplined translation between them, guided by boundary models and contextual awareness.

By maintaining this discipline, inquiry remains coherent without becoming restrictive. It allows multiple perspectives to coexist, provided each is applied where it is meaningful. In this way, clarity is achieved not by eliminating complexity, but by respecting the conceptual structures through which understanding is possible.

Appendix F.6—Integrating Vedanta And Physics (Methodological discipline)

The integration of Vedantic thought and modern physics in this inquiry is not pursued as a synthesis of doctrines, nor as an attempt to validate one tradition through the authority of the other. Instead, it is approached as a **methodological dialogue**—one that allows distinct traditions to illuminate complementary aspects of inquiry while preserving their respective domains and constraints.

Vedanta offers a long-standing discipline of questioning centered on perception, illusion, selfhood, and the limits of conceptual knowledge. Physics, by contrast, advances through mathematical formalism, empirical validation, and predictive coherence. Each tradition operates under different epistemic rules, and meaningful engagement between them requires respecting these differences rather than collapsing them.

This inquiry draws selectively from Vedantic insights where they function as **conceptual orienting tools**, particularly in questions concerning observer-dependence, appearance versus underlying structure, and the limits of descriptive language. Such use does not treat Vedantic concepts as empirical claims, nor does it reinterpret physical theories as metaphysical affirmations.

Methodological discipline is maintained by applying each framework **only within the scope where it is meaningful**. Physics addresses measurable structure and behaviour; Vedanta addresses experiential inquiry and conceptual limits. Where parallels appear, they are treated as resonances rather than equivalences.

By sustaining this disciplined separation alongside thoughtful dialogue, the inquiry benefits from both traditions without subordinating one to the other. Integration here does not mean unification, but **coherent coexistence**—allowing multiple ways of knowing to inform understanding while guarding against category errors, overreach, and metaphysical closure.

Appendix F.7—Critical Filters for Evaluating Scientific Claims

In an era of rapid discovery and accelerated dissemination, scientific claims often circulate faster than their careful evaluation. This inquiry therefore applies a set of **critical filters**—not to dismiss new ideas, but to distinguish robust insight from premature interpretation, speculation, or overreach.

The first filter is **domain alignment**. A claim must be evaluated within the domain it addresses. Empirical findings, theoretical models, philosophical interpretations, and metaphysical extrapolations each operate under different standards of validity. Confusion arises when results from one domain are presented as conclusions in another without appropriate justification.

The second filter is **evidentiary grounding**. Claims should be assessed based on the quality, scope, and reproducibility of the evidence supporting them. Singular results, unreplicated experiments, or indirect inferences may be suggestive, but they do not carry the same weight as convergent lines of evidence.

The third filter concerns **interpretive restraint**. Even well-supported findings can be overextended through metaphor, narrative framing, or speculative extrapolation. This inquiry favours interpretations that remain close to what the evidence actually warrants, resisting the temptation to draw existential or cosmological conclusions beyond the data.

A fourth filter is **conceptual coherence**. New claims must be examined for internal consistency and compatibility with established frameworks, or else clearly articulate where and why existing frameworks may be insufficient. Disruption alone is not a virtue; meaningful progress preserves coherence even while revising understanding.

Together, these filters support a mode of inquiry that is neither sceptical to the point of paralysis nor receptive to the point of credulity. They enable openness without loss of rigour, allowing new ideas to be explored thoughtfully while safeguarding against confusion, hype, and metaphysical closure.

Appendix F.8—Zero, Nothingness, And Emptiness: Conceptual Clarifications

Concepts such as *zero*, *nothingness*, and *emptiness* are often treated as interchangeable in casual discourse, yet they belong to **distinct conceptual domains**. Confusing them leads to philosophical overreach, metaphysical speculation, and category errors—particularly when scientific language is stretched beyond its proper scope.

Zero is a formal construct. In mathematics and physics, it denotes a reference value within a defined system—such as the absence of quantity, charge neutrality, or a baseline for measurement. Zero is not the absence of structure; it presupposes a framework within which it is defined and meaningful.

Nothingness, by contrast, is a metaphysical concept. It refers to the absence of entities, properties, or distinctions altogether. Science does not operate on nothingness in this sense, because any observation, measurement, or description already presupposes existence and structure. Attempts to treat physical vacua or ground states as "nothing" therefore commit a category error.

Emptiness, particularly as understood in philosophical traditions such as Vedanta and Buddhism, occupies a different register altogether. It does not denote non-existence, but rather the absence of inherent, independent essence. Emptiness points to relational dependence and structural contingency—not to a void devoid of reality.

In this inquiry, these distinctions are maintained with care. Physical theories may describe low-energy states, vacuum fluctuations, or symmetry ground levels, but such descriptions remain **within structured systems**. They do not speak to absolute nothingness. Likewise, philosophical notions of emptiness are treated as conceptual orientations, not empirical claims.

Clarifying these distinctions helps prevent misplaced questions—such as asking what "came before" structure, or whether physics explains why there is "something rather than nothing." Such questions may be meaningful philosophically, but they do not belong within the explanatory reach of physical theory.

By maintaining this discipline, the inquiry avoids metaphysical inflation while preserving conceptual clarity. It allows scientific, mathematical, and philosophical ideas to coexist without collapse—each operating where it is meaningful, and no further.

Appendix F.9 —Consciousness As A Distributed Phenomenon

Recent work in neuroscience increasingly suggests that consciousness may not arise from any single region of the brain, but from patterns of interaction distributed across neural networks. Rather than being produced as a localised output, conscious experience appears to correlate with the integration and coordination of information across the system.

This perspective reflects a broader pattern observed throughout this inquiry: that what appears as a "thing" is often better understood as a stable pattern emerging from relations.

In this sense, the brain may be less a generator of consciousness and more a constraint structure within which conscious experience becomes possible.

This interpretation remains an active area of research. Its significance here lies not in its finality, but in the continuity it reveals—that across domains, what appears as substance often resolves into structure, and what appears as origin into relation.

F.10 —Structure, Relation, And Constraint: A Relational Methodological Stance

Throughout this inquiry, reality is approached not as a collection of independent objects, nor as a construct arising solely from perception, but as a structured field of relations. What exists, persists, and becomes observable is shaped by constraints—physical, informational, and contextual—that govern how relations can form and evolve.

This perspective does not adopt the classical opposition between subjective and objective reality; instead, it treats reality as relational, arising through interaction rather than from perception alone or isolated objects.

Methodologically, this stance prioritises coherence over classification, structure over assertion, and explanatory discipline over speculative excess. It seeks understanding without reductionism, avoids category errors, and resists the temptation to force metaphysical predictive closure.

Appendix G—Beyond the Observable: Depth, Liberation, and Understanding

At the limits of observation, inquiry turns inward—toward structure, understanding, and freedom.

G.1 Layers Beneath Observation: The Question of Depth

G.2 Cycles, Freedom, and the Question of Liberation

G.3 From Observation to Understanding: A Cognitive Ascent

G.4 Ignorance, Comfort, and the Role of Inquiry

G.5 Structure Without Finality

G.6 Constraint, Form, and Emergence: An Everyday Illustration

Appendix G.1— Layers Beneath Observation: The Question Of Depth

In the course of this inquiry, reality has been approached through multiple descriptive regimes—classical, quantum, biological, and cognitive. Each of these domains reveals structure, but none can claim finality.

Modern physics suggests that even the quantum domain may not be fundamental. Below the Planck scale—beyond current experimental access—may lie pre-geometric or informational structures from which spacetime itself emerges.

Speculative extensions of this idea appear indirectly in what is sometimes referred to as the "zeptouniverse," where extremely small scales and high-energy phenomena hint at deeper organising principles. These do not constitute separate "layers" in a spatial sense, but rather regimes of description governed by different constraints.

What appears as matter, field, or particle may therefore be understood as an emergent manifestation of deeper relational structure.

This perspective is consistent with the methodological stance adopted in this work:

> Reality is not treated as a collection of objects
>
> Nor as a construct of perception alone
>
> But as a structured field of relations governed by constraint

At deeper levels, what changes is not reality itself, but the language required to describe it.

Appendix G.2—Cycles, Freedom, And The Question Of Liberation

Scientific cosmology allows for the possibility of cyclic universes—expansion, contraction, and renewal across vast timescales. These cycles describe the evolution of matter, energy, and spacetime.

Philosophical traditions, particularly in Vedanta and related schools, describe a different form of cyclicity—that of experience, identity, and rebirth (samsara).

It is important, methodologically, not to conflate these domains.

- Cosmological cycles belong to physical description
- Existential cycles belong to experiential and philosophical interpretation

Within these traditions, liberation (moksha) does not imply escape from the Universe as a physical system, but release from the cycle of identification with transient forms.

While science does not posit such liberation as an empirical category, certain analogies may be noted cautiously:

- Systems can transition between states under changing constraints
- Behaviour can shift qualitatively when underlying parameters cross thresholds
- Stability and instability emerge from structural conditions, not external imposition

In this limited sense, liberation may be understood—metaphorically—as a transition in the mode of identification, rather than a movement in space or time.

This inquiry does not assert such interpretations. It acknowledges them as part of a broader human attempt to understand the relation between structure, experience, and freedom.

Appendix G.3—From Observation to Understanding: A Cognitive Ascent

Across the chapters of this work, a pattern emerges—not only in the structure of reality, but in the structure of understanding itself.

A useful way to frame this progression is as follows:

Observation organises sensory input

Perception organises observation into patterns

Awareness brings attention to those patterns

Knowledge structures and communicates them

Wisdom integrates knowledge with context, consequence, and judgement

Each stage does not replace the previous one; it reorganises it.

Scientific inquiry operates primarily within observation and knowledge. Philosophical inquiry extends into awareness and interpretation. Lived experience tests the transition toward wisdom.

A critical distinction follows:

Knowledge can be accumulated

Wisdom must be integrated

Without awareness, perception remains mechanical. Without structure, knowledge remains fragmented. Without judgement, knowledge may become misapplied.

In this sense, the inquiry presented in this book is not merely an exploration of external reality, but a movement within this cognitive ascent—from description toward understanding.

Appendix G.4—Ignorance, Comfort, And The Role Of Inquiry

It may appear that not knowing provides a form of ease. Questions unsettle; inquiry disrupts familiar structures of thought.

However, the apparent comfort of ignorance arises not from stability, but from the absence of contradiction.

When deeper structure is encountered—whether in science, philosophy, or lived experience—that comfort gives way to a more demanding form of coherence.

A useful distinction may be drawn here.

What is often described as "bliss in ignorance" is not a form of understanding, but a temporary absence of conflict. It resembles stability only because contradiction has not yet been encountered.

In that sense, ignorance does not produce clarity; it postpones it.

The movement toward knowledge, therefore, is not driven merely by curiosity, but by an implicit recognition that borrowed coherence cannot sustain itself indefinitely.

This does not imply that knowledge eliminates discomfort. On the contrary, it often introduces it. But it replaces fragile comfort with a more stable form of understanding.

In this limited sense, what appears as bliss in ignorance may be closer to suspension than resolution—a form of cognitive anaesthesia rather than clarity.

The purpose of inquiry, therefore, is not to replace discomfort with certainty, but to replace ambiguity with clarity.

In this sense:

Ignorance does not provide understanding

Knowledge does not guarantee coherence

Inquiry is the process through which coherence is approached

This work aligns with that orientation. It does not seek to conclude, but to clarify.

Appendix G.5—Structure Without Finality

This appendix extends, but does not alter, the central methodological position of the book:

> Reality is approached through structure, relation, and constraint
>
> Descriptions are valid within regimes, not across all scales
>
> Scientific and philosophical languages are complementary, not interchangeable

At its limits, inquiry encounters questions that may be structurally ill-posed within a given framework. Recognising such limits is not a failure of understanding, but a refinement of it.

What remains, then, is not a final answer, but a stable orientation:

to observe carefully, to reason with discipline, to integrate without overreach, and to remain open where closure is not yet warranted.

Appendix G.6—Constraint, Form, And Emergence: An Everyday Illustration

The following illustration is not a simplification, but a way of making the underlying structure more visible.

The central idea of this work may be illustrated through a simple observation.

Form does not arise independently. It emerges when constraints act upon possibility.

Consider flowing water. By itself, it has no fixed shape; it moves freely without stable form.

When it enters a depression in the ground, it takes the shape of that boundary. When placed in a vessel, it assumes the form defined by its container. When temperature decreases sufficiently, it transitions into a structured solid.

In each case, the substance remains the same, but the form changes in response to constraint.

Without constraint, there is flow but no stable structure. With constraint, relations are stabilised, and form becomes observable.

This illustrates a more general principle:

Constraints do not merely limit what is possible; they enable form to emerge.

In this sense:

Constraints shape relations

Relations give rise to structure

Structure enables emergence

What appears as form is not imposed from outside, nor inherent in isolation. It is the outcome of interaction within constraint.

Poetic Postscript—Two Mirrors Of The Mind

As a final reflection to this inquiry, the following poem offers a lyrical summation of the book's central theme: the interplay between fundamental science and emerging forms of intelligence, as two mirrors through which we glimpse deeper layers of reality.

Two Mirrors of the Mind

In ancient caves, with fire in hand,
We shaped a world we didn't understand.

Now circuits hum and qubits spin,
A new kind of magic stirs within.

One mirror shows the cosmic thread,
Where particles dance long thought dead.
A quantum sight—uncertain, true—
Where logic bends and light breaks through.

The other breathes a dream made real,
Where minds are mirrored, taught to feel.
Neurons echoed in silicon's glow,
Learning truths we scarcely know.

We harvest waves we cannot see,
And teach machines to mirror us.
Yet in our hearts, a question grows:
Do we create what we don't yet know?

Through tangled phase and weighted node,
We walk a dimly lit new road.

With reason faint and wonder wide,
We ride the storm, not just the tide.

So here we stand—no gods, no guide—
With quantum thought and neural pride.
Two mirrors held to space and soul—
One seeks the truth, the other whole.

Epilogue—Approaching Truth

Every inquiry must end not with certainty, but with a deeper respect for what remains unresolved.

This book began by asking questions about space, time, matter, life, and consciousness. Along the way, it explored symmetry and asymmetry, emergence and impermanence, intelligence and interdependence. Yet none of these explorations were meant to close debate or declare final answers. They were meant to trace coherence—to see what holds together when ideas from different domains are placed in honest conversation.

If there is a quiet lesson that runs through these pages, it is this: reality reveals itself not all at once, but through structure, constraint, and relationship.

Truth as an Approach, Not a Possession

Truth, as it appears here, is not a statement to be defended. It is not a doctrine, nor a destination.

Truth is what survives inquiry.

In science, truth is provisional—refined by evidence, corrected through error, strengthened by falsification. In philosophy, truth is approached through reason, consistency, and clarity. In lived experience, truth reveals itself through alignment—when understanding, action, and consequence resonate.

Across these domains, a common pattern emerges: truth is not seized; it is approached asymptotically.

What endures is not certainty, but coherence. What matters is not finality, but fidelity to inquiry.

The Human Place in the Inquiry

The Universe explored in this book is not a detached object. It is the very condition that allows questioning to arise.

Stars forged the elements of the body. Fields and forces shaped matter into form. Life organised itself into awareness. Awareness learned to reflect upon itself.

To inquire into reality is therefore not an act performed from outside. It is reality examining itself through human attention.

This does not elevate humanity above nature. It places humanity within nature's unfolding—as a temporary, thoughtful participant.

On Knowledge, Humility, and Faith

Knowledge collects descriptions. Understanding seeks the constraints that make those descriptions necessary.

Every serious inquiry encounters uncertainty. Every honest thinker meets the limits of explanation.

At those boundaries, two responses are possible: one of fear, and one of humility.

Humility does not abandon reason; it protects it. It allows inquiry to continue without collapse into dogma or despair.

In that spirit, one may say:

Faith is the quantum error correction of the soul—maintaining coherence against the decohering noise of fear and doubt.

Faith here is not belief without evidence, but trust in the process of inquiry itself—that careful questioning, disciplined thinking, and openness to revision are worthwhile, even when outcomes are uncertain.

Intelligence and the Expanding Horizon

As intelligence—human and artificial—extends its reach, new patterns may become visible that no single mind could hold alone.

Yet recognition is not awareness. Computation is not comprehension.

The tools we build may help us see further, but they do not replace the responsibility to understand.

The deeper question therefore remains: not what intelligence can reveal, but how wisely it is interpreted.

A Closing Reflection

Every animal leaves traces of what it was. Human beings leave traces of what they create.

Civilisations rise through what they build—but endure through what they understand.

Across life, matter, and cosmos, a pattern quietly repeats. What appears as diversity is often variation within constraint. What appears as complexity is often structure seen across scale. What appears as mystery is sometimes the limit of our descriptive tools.

The same reality expresses itself differently—as biology, as physics, as consciousness, as cosmos. Not as separate domains, but as interconnected regimes of description.

We do not stand outside this reality. We participate in it.

Final Convergence

We do not approach truth by possession, but by coherence of understanding. We do not complete inquiry by certainty, but by humility before what remains unknown.

A subtle limitation of human imagination is to see reality as static, when it is in fact dynamic—an ongoing process of change and relation.

In a cosmos that exceeds our comprehension, the wisest stance is not mastery, but sustained wonder. What remains is not finality, but sādhana—a disciplined openness to deeper structure. And in that openness, the knower, the known, and the knowing are held in quiet alignment.

Final Thought

Everyone thinks of changing the world, but no one thinks of changing himself.

—Leo Tolstoy

About The Author

K V V S Murthy *(Satyanarayanamurthy Veera Venkata Korlepara)*

K VV S Murthy is a lifelong learner whose professional journey spans banking, freelance journalism, economics, finance and accounts, learning and development, strategy, and systems thinking across diverse organisational and institutional settings.

Trained in analytical disciplines yet drawn equally to philosophical and scientific questions, he has spent decades exploring the intersection of numbers and meaning—where structure meets uncertainty and precision coexists with humility.

This book does not arise from academic specialisation in physics or metaphysics, but from sustained curiosity—a belief that reality is best approached through questions that cross disciplines rather than answers confined within them.

An Inquiry into the Nature of Reality reflects this journey: not as a declaration of truth, but as an invitation to think more carefully, observe more honestly, and remain open to what may yet be understood.

Connect with the author: linkedin.com/in/satyanarayanamurthy-korlepara-16aa4032

Acknowledgements

This work grew out of a lifetime of questioning rather than a single discipline or moment. I acknowledge, with gratitude, the many visible and invisible teachers along the way—scientists who dared to doubt, philosophers who refused easy answers, sages who spoke in paradox, and readers who believe that inquiry itself is a form of reverence.

I also acknowledge the role of artificial intelligence as a contemporary instrument of exploration—not as an authority, but as a mirror that sharpened questions, tested coherence, and accelerated reflection.

Above all, I acknowledge inquiry itself—the quiet force that keeps curiosity alive long after certainty fades.

I am grateful to my family for their quiet support and understanding throughout this journey.

www.ingramcontent.com/pod-product-compliance
Lightning Source LLC
LaVergne TN
LVHW100255210926
848480LV00003B/926
* 9 7 9 8 9 0 4 5 1 1 4 3 2 *